The human population of the world is increasing by about 1.5% per annum, adding about one hundred million people to the human ark each year. Not all agree that population growth on this scale constitutes a problem, but there is wide acceptance that the world's human population cannot go on growing indefinitely. Where do the limits lie, and how can they be determined? What are the problems caused by population growth and how can we safeguard the future of our planet? In this important new book, leading authorities examine the implications of rapid human population growth for global stability and security. Avoiding the hysteria and over-statement that so often characterize discussions of human population issues, the book represents an important assessment of current prospects for the process of sustainable development, based on care for the environment.

Nicholas Polunin died shortly before the publication of this book. At the time of his death he was President and Chief Executive Officer of the Foundation for Environmental Conservation, Geneva, Switzerland.

DEDICATED
with Loving Admiration
of her unfailing Loyalty and Patience
to my Darling Wife

HELEN EUGENIE POLUNIN

who has effectively shared my Problems
and encouraged my Aspirations
for Humanity and Nature
through already a half-century of
Happy and Productive Marriage

POPULATION AND GLOBAL SECURITY

Edited by

NICHOLAS POLUNIN

CAMBRIDGE
UNIVERSITY PRESS

CAMBRIDGE UNIVERSITY PRESS
Cambridge, New York, Melbourne, Madrid, Cape Town, Singapore,
São Paulo, Delhi, Dubai, Tokyo, Mexico City

Cambridge University Press
The Edinburgh Building, Cambridge CB2 8RU, UK

Published in the United States of America by Cambridge University Press, New York

www.cambridge.org
Information on this title: www.cambridge.org/9780521635394

First published 1998

A catalogue record for this publication is available from the British Library

ISBN 978-0-521-56372-7 Hardback
ISBN 978-0-521-63539-4 Paperback

Contents

Nicholas Polunin
1909–1997

Professor Dr Nicholas Polunin, as he liked to be styled in continental fashion, who has died aged 88, was a distinguished scientist in many more fields than his first speciality of botany. He explored the Canadian Arctic and Greenland and discovered the last major islands to be added to the world map. He also spent his last 20 years in spirited support of the environmental cause, demonstrating the same elegant élan he brought to his academic endeavours.

For all that he became an environmentalist who prided himself on being abreast of the latest advances in analysis and evaluation, he retained certain traits of the old school. It was this traditionalist side that, in the view of many colleagues, bespoke the core Polunin: he was always his own person. For instance, he could not abide the use of "billion" in its American sense, and writers in his journals were obliged to speak of billion only when they meant one million million, otherwise they had to stick with 1,000 million.

Still more illuminatingly, he eschewed the use of "ecosystem" when speaking of the planetary biosphere and its physical underpinnings. He insisted that an ecosystem should have specific boundaries, as in the case of a forest or an island or a puddle. How, he would ask, laughing at the absurdity of the notion, could Australia's ecosystem extend across the seas to be part of a global entity? Not all his peers would agree with him, but he persisted with his views in disregard for whatever "fashionable science" might prefer. It was a characteristic that bothered some and endeared him to many.

In these and other significant respects, Polunin was occasionally out of step with his times, and he minded not a jot. His dissident views reflected a passion for that prized scientific virtue, rigour, which implies a strictly disciplined approach to the scientific enterprise. Polunin displayed rigour in all his professional activities: precision and accuracy were all. He was sometimes puzzled when his colleagues did not display a parallel passion. And he extended this passion to another of his great loves, the English language. It was a delight to watch him savour a phrase, or to test a sentence as if tasting it.

Nicholas Polunin was born in Checkendon, Oxfordshire, the son of a Russian father and an English mother. In 1932 he graduated from Christ Church, Oxford, with a first in botany and ecology (an innovative combination in those days). He went on to a Master's at both Yale and Oxford, before finishing up with a DPhil and a DSc at Oxford. He then spent some time as a research associate at Harvard. Along the way, he pursued research on arctic and sub-arctic flora, travelling widely in Spitsbergen (which he crossed alone), Lapland, Iceland, Greenland, Labrador, Hudson Bay, Akpatok and Baffin Island, leaving plant collections in the Canada Museum, the Harvard Herbarium, the British Museum, and the Fielding Museum at Oxford. From 1932 to 1947, he was botanical lecturer and tutor in several Oxford colleges, especially New College, and for five years from 1947 he was Professor of Botany at McGill University in Montreal.

Thereafter Polunin expanded his professional purview by heading for the tropics. In the late 1950s he participated in the establishment of the University of Baghdad, and in the early 1960s he performed a similar function with the University of Ife, Nigeria, where he actually planned the new campus as well as serving as head of the science faculty. In later life he held academic appointments on four continents.

Among his notable scientific achievements was the discovery in 1936 of plants in south-western Greenland introduced by the Vikings. Still more remarkably, in 1946 Polunin the explorer discovered two islands located in the Foxe Basin to the west of Baffin Island and subsequently named Prince Charles Island and Air Force Island. Even more exceptional was his demonstration of the summertime existence of microbial spores and pollen grains high above the North Pole. He also tracked down several new species of arctic plants.

Although he retained his British nationality, he moved to Switzerland in 1959 by invitation, and from his Geneva base embarked on what many consider were the prime accomplishments of a highly diverse career. From the start he had espoused the cause of holistic science – an exceptionally pioneering departure in the 1930s, since even today it is viewed askance by many in the citadels of single-discipline science. This expansive approach led Polunin to be one of the first scientists to ask whether our environments, and especially wild life habitats, were threatened by the rise in human numbers and human demands. A conservationist by personal inclination as well as by professional calling, he became the founding editor of the journal, Biological Conservation in 1967, until he moved on in 1974 to play a similar role in an even more prestigious journal, Environmental Conservation, which now enjoys circulation and acclaim in dozens of countries.

At the same stage in his still productive life (he was then in his mid-sixties), he established the Foundation for Environmental Conservation, under whose auspices he organised four international conferences in Helsinki, Reykjavik, Edinburgh and Budapest. Each of these conferences led to first-rate books reviewing environmental prospects and highlighting both the problems and opportunities implicit therein. Notable among these books were *Maintenance of the Biosphere* (1990) and *Surviving with the Biosphere* (1993). A final fine book was *Population and Global Security* (1994), revised in his last year.

Polunin's efforts were recognised through numerous awards: the US Order of Polaris (twice), the Ramdeo Medal of India, the Marie-Victorin Medal of Canada, the Sasakawa Environmental Prize of the United Nations, the President Quo Mo-Jo Medal of China, the Vernadsky Medal of the Soviet Union, the Zechenyi Medal of Hungary, the Order of the Golden Ark of the World Wildlife Fund, and the Global 500 Roll of Honour of the UN Environment Programme. In addition, several geological features in the Canadian Eastern Arctic were named after him, as were a number of Canadian Arctic and Greenlandic plants.

His work is well summarised by a recreation (sic) item listed in his entry in *Who's Who*: "working towards establishing a planetary eco-network of environmental conservational watchdogs". The same entry mentions that he was "embarrassed by personal publicity but welcomes news media support provided it is accurate and dignified". This is Polunin exactly. He crusaded for the environment with a rigour and vigour that placed him among the most worthy, and he was already in the van when most of today's crusaders were still in school. At the same time, he brought to the cause a courtly spirit that often set him apart. I especially appreciated him for his gentlemanly demeanour, however frantic the fray, and whatever the different deportment of those around him. Of all crusaders, he was a verray parfit gentil knight.

His first wife Helen died in 1973, whereupon he married a second Helen, who survives him, as do his two sons and one daughter.

Norman Myers

First published in *The Guardian* newspaper on 13th December 1997. Reproduced with permission.

Preface

The impact of human activities, with our pandominant species' still-increasing numbers and profligacy, is becoming ever-more evident to enlightened scientists and threatening to our unique Planet Earth. Consequently it is of crucial importance that the extreme gravity of this situation be recognized by all responsible people. For with proper discipline and personal restraint in the matter of procreation, this most alarming threat can be countered at least sufficiently to avoid the worst foreseeable ecodisasters — such as decimation of the stratospheric ozone shield, adverse climatic change, and famine resulting from insufficiency of cultivable land and of edible freshwater and marine resources. Yet continuing along the line of the detailed General Introduction to our Limited Geneva Edition[1] preceding the present work, we may safely say that, to some originally small but ever-increasing extent practically from its inception, Humankind has 'always' been in conflict with the rest of evolved Nature — however one may view that last, undefined but to us marvellous, phenomenon[2].

Realizing some years ago that Humankind's pandominance and profligacy were bringing it 'clearly on a collision course with Nature' and hence with its own long-term future, we decided on the need of an anthology which would treat important aspects of the extremely grave predicament most effectively through the writings of chosen leaders, and so produced the Foundation for Environmental Conservation's Limited Geneva Edition[1] of this work particularly as background reading for participants of the United Nations' International Conference on Population and Development, which was held in Cairo, Egypt, in September 1994[3].

Being commissioned subsequently by Cambridge University Press to edit an updated, 'post-Cairo', version of that book, the undersigned had hoped this could be done while the 'spirit of Cairo'[3] remained fresh in people's minds; but he had first to edit the current and next (his last of 22) annual volumes of the Foundation for Environmental Conservation's international quarterly Journal *Environmental Conservation* before transferring its publication to Cambridge University Press and then prepare for printing the first edition of the Foundation's *World Who Is Who and Does What in Environment & Conservation*[4], obtain and edit two needed new chapters for this post-Cairo anthology, and finally write two long-promised chapters for festschrifts of eminent friends, so that only late in 1996 was he able to get down to editing this CUP book. These delays have necessitated considerable updating as well as revision on the part of contributors, to whom readers and all others concerned — but particularly the undersigned — are accordingly indebted.

Nicholas Polunin, *President & CEO*
The Foundation for Environmental Conservation
7 Chemin Taverney (7th & 8th floors)
1218 Grand-Saconnex
Geneva, Switzerland

---

[1] *Population and Global Security* (Environmental Challenges II) edited by Nicholas Polunin & Mohammad Nazim. (Limited Geneva Edition for the United Nations Population Fund [UNFPA] *et al.*). Published by the Foundation for Enviromental Conservation, Geneva, Switzerland: xi + 285 pp., illustr. & tables, 1994.

[2] Note here that we avoid possible conflict by according our species a general common English name embracing women as well as men, that we put 'always' in quotes (*sic*) to avoid controversy with geneticists and theologians who were not present at our species' beginning, and that we place 'the rest of' before 'evolved Nature' as we have long been convinced that Humankind has to be treated as part of Nature (and in turn so treat itself) if our world is to have a fair chance of long-term equable continuation, let alone lasting survival.

[3] Guest Comment: Reflections on the UN's International Conference on Population and Development, by Dr Pramilla Senanayake. *Environmental Conservation*, 22(1), pp. 4–5, 1995.

[4] *World Who Is Who and Does What in Environment & Conservation*, Edited by Nicholas Polunin, Complied by Lynn M. Curme. Published for the Foundation for Environmental Conservation by Earthscan, 120 Pentonville Road, London N1 9JN, England, UK: viii + 592 = 600 pp., January 1997. This First Edition contains more than 1,300 biographies as well as detailed Appendixes enabling users readily to identify, locate, and contact, experts in any geographical region and/or pertinent field of need.

1

Population Growth and Global Stability

by

NAFIS SADIK, MD (Karachi)

Executive Director,
United Nations Population Fund (UNFPA),
220 East 42nd Street, New York,
NY 10017, USA;
Secretary-General of ICPD

INTRODUCTION

Termination of the 'cold war' brought a sigh of relief around the globe and spawned new hopes for a more tranquil and safer world order than the one which had dominated international affairs since the late 1940s. Unfortunately these hopes would seem now to have given way to a feeling of dejection. However, the disappointment occurs not, as is sometimes argued, because new forms of violence and conflict have replaced those that were regarded as typical progeny of the cold war. Rather, the disappointment appears to stem from a failure to distinguish clearly between the many forms of conflict and violence which had their roots in the cold war and those that did not. There was a tendency to attribute many or almost all ills to the existence of the cold war, the disappearance of which would therefore surely improve matters all around. The ethnic and nationalist conflicts that are now seen by many as symbols of the post-cold war 'disorder' are, however, but tragic examples of a breed of conflicts which has roots going back for centuries. Many existed during the cold war itself. To these have been added those that were unleashed by the collapse of centralized economies.

Many features of today's or very recent conflicts — whether in the Balkans, Afghanistan, the Caucasus, Rwanda, Somalia, Zaire, or elsewhere — are all-too-familiar, and enable us, therefore, to recognize them for what they are, namely ethnic, religious, or economic. However, there are other features and signs which, though not entirely new, are much less familiar, as they were much less conspicuous in the past than they are now, and moreover than they

1

are likely to be in the future. Most alarming among these is the rapid growth of the world's human population and the implications this may have for global stability and security.

Discussed below are some of the consequences which are already visible or may result from a failure to stabilize the world's population. While some of these consequences may be unavoidable, the possible effects of others may be ameliorated and even prevented by creating conditions in which women and men are empowered to make their own choices in matters related to family formation and child-bearing.

In that last respect, probably no international conference has provided better guidelines and more widely-supported recommendations than the International Conference on Population and Development (ICPD), held in Cairo in September 1994 (*cf.* Senanayake, 1995). The ICPD brought to the fore a new paradigm of population and development — one which puts people first, emphasizes the importance of empowering women, and which places reproductive health, family planning, and sexual health, at the top of the population agenda. What is essential to understand is that in this approach, a balance between population and resources can only be achieved when individuals are empowered to make their own choices, and when men and women are provided with the information and means to carry out their decisions.

It is also important to stress that, although the focus of this chapter is on population growth, we should all along be mindful of the fact that many factors are at work and influence one another. The actual situation is perhaps best captured in this quotation from the ICPD Programme of Action (1994).

> 'There is general agreement that persistent widespread poverty as well as serious social and gender inequities have significant influences on, and are in turn influenced by, demographic parameters such as population growth, structure, and distribution. There is also general agreement that unsustainable consumption and production patterns are contributing to the unsustainable use of natural resources and environmental degradation as well as to the reinforcement of social inequities and of poverty with the above-mentioned consequences for demographic parameters.' (United Nations, 1995).

ON THE NOTION OF GLOBAL STABILITY

Something — be it a structure, a government, a state, an international order, etc. — is stable or exhibits *stability* when it is 'firmly fixed or established; not easily adjusted, destroyed, or altered', according to *The Concise Oxford Dictionary of Current English* (9th edn, 1995).

Stability clearly implies a reasonably lasting state of affairs; there is no such thing as transient stability. Stability is usually also seen as a positive quality because it implies predictability, without which no social life or ordered human interaction would be possible.

Although 'fixed or established', stability should not be confused with rigidity and immobility. Only vital structures which possess at least a modicum of elasticity and adaptability will last. Thus, whereas some types of change, and perhaps even the rapidity with which change comes about, may have a corrosive effect on stability, other types of change will not. So as not to undermine stability, however, contrived change, in particular, must normally be carried out according to some rules which are generally accepted within the structure in question.

Stability may be a feature of different types or spheres of activity — political, economic, and/or social. Stability may also be a characteristic of, for example, local, national, and international, levels. The discussion in this essay centres on the effects of population growth on political stability at the national and international levels.

In spatial terms, stability may be partial or complete. It is doubtful whether the modern world has ever experienced complete stability in the sense of its having been globally pervasive *quite evenly*. It certainly does not do so today. Stability ebbs and flows like the sea, but it does so according to laws or sets of circumstances which are only imperfectly understood. There are always, at any time, one or more countries which are experiencing instability and disorder.

A most important consideration concerns the relationship between stability and the use of force. The two are often seen as related in much the same way as fire and water are; the presence of one signals the absence of the other. Everyday experience, however, tells us that a political system may still be deemed stable even though a degree of force is being used — as, for example, in the case of a legally established authority seeking to maintain civil order. In the international system, the use of force has never been wholly frowned upon by the community of states, mainly because force may have to be used in self-defence or against powers that try to overturn or subvert the system. In both cases, stability is a function of the degree of force, and also of the observance of certain rules regarding the use of force — especially in civil society.

Our point here is that the presence of force and violence does not necessarily signal the absence of stability. But neither, of course, are all levels of force compatible with the presence of stability. The level at which force is sufficient to destroy all claims to stability will vary from one place and society to another. It is not the intention in this essay, however, to take on the impossible task of predicting exactly whether, and if so when and where, this will occur.

POPULATION GROWTH: A PROBLEM?

According to the United Nations' latest revision of its *World Population Prospects*, fertility rates are declining more rapidly in most regions of the world than had been anticipated. As a result, the global population is now growing, on average, by 1.5% a year — down from the 2% growth-rates of the 1960s and 1970s. In a number of developing countries, however, growth-rates persist at 2.5% or even more. And because of the huge population base, we are adding one thousand million people to the human ark every 12 years until AD 2011 — the fastest increase in human numbers ever recorded. According to the UN's revised medium-range population projections, the Earth will have six thousand million people in AD 1999, seven thousand millions by AD 2011, and eight thousand millions in AD 2025. The global population is expected to keep on growing until around the end of the next century, when it should stabilize at close to 10.7 thousand millions — about double the current world population.

The magnitude of the problem is more easily understood by looking at the demographic circumstances of various parts of the developing world, in particular. Of the projected growth in population, 95% is expected to take place in Africa, Asia, and Latin America. Africa's population is growing at 2.9% a year, the fastest ever for an entire region. Eastern and western Africa are growing at 3.1%, Western Asia is growing at 2.8% a year, and Southern Asia (the Indian subcontinent) is growing at 2.2%. Asia overall is growing at 1.8%, and Latin America at nearly 2%.

By the year 2025, Asia, with a population of roughly 3.2 thousand millions in AD 1992, is expected to have 4.9 thousand millions; Africa, with 682 millions, is expected to more than double to 1.58 thousand millions; and Latin America, with 457 millions, is projected to reach 701 millions.

Starker still is the predicament of many individual countries. Already densely populated, Bangladesh is projected to nearly double its population between the years 1990 and 2025. Numerous countries in Africa and the Middle East, among them Angola, Congo, Iraq, Jordan, Kenya, Libya, Madagascar, Malawi, Niger, Nigeria, Rwanda, Saudi Arabia, Syria, Tanzania, and Zaire, have a population growth-rate well above 3% a year and are projected to more than double to almost treble their populations in the course of a generation or so. Several of these are among the poorest countries in the world, and many already suffer severe shortages of resources as essential as, for example, clean water.

Not all agree that population growth on this scale constitutes a problem, let alone a menace. Some argue that the Earth can accommodate many times more people than today's population, and that the present emphasis on slowing population growth is, therefore, misguided.

All do agree, however, that the world's human population cannot go on growing indefinitely. The question, therefore, is not whether there are limits to how many people the Earth can sustain, but rather where the limits lie and how they should be determined. Many believe that, for example, a doubling of today's numbers could well be too many. The fact is that the margin of error as to how many people the Earth can sustain is not *known* even approximately. The number depends not only on the availability and rate of exploitation of resources, which can probably be quantified and measured with some degree of accuracy, but also on a number of factors that essentially concern the style and quality of life that all people might wish to enjoy, in addition to a number of other factors to which much uncertainty is attached. This is an important reason why many are convinced that it is better to err on the side of caution by adopting the approach recommended, *inter alia*, in Principle 15 of the Rio Declaration on Environment and Development.

Let us, nevertheless, accept for a moment the view that, from a global perspective, the Earth's environment and resources can accommodate more, even many more, people than they do today. Accepting that view does not mean, however, that we are not now faced with a population growth problem and will not continue to face one in the future. The growth problem arises because we do not live in a world with perfect mobility of resources and people, in a world in which no impediments stand in the way of people moving from too-densely-populated lands to lands where resources are relatively more plentiful — in other words, in a world which encourages a rational balance between people and resources.

The world we live in is divided between different races and countless clans, tribes, ethnic groups, nations, and states. The *raison d'être* for each of these groups is based upon claims to being clearly distinguishable from others. Their defining characteristics are not cultural and other similarities so much as contrasts and differences, which become objects of veneration and preservation and in turn provide justification for the exclusion of others. There are also important political, economic, and social, barriers, and an uneven distribution of natural resources between countries, regions, and continents. Institutional inefficiencies and a lack of public and private capabilities in many, if not most, fields are familiar aspects of everyday life — particularly in the less-developed countries.

At both national and international levels, obstacles to the achievement of a balance between people and resources are sometimes formidable. It is not unusual to find, in parts of a country, large

numbers of undernourished and even starving people due to shortages, whereas, in other parts of the same country, there is excess
supply and food is exported abroad. At the international level, for
years it has been maintained that there is no global food-shortage.
The starvation or undernourishment of hundreds of millions of
people has been blamed on 'distribution problems'. But what are
called distribution problems as often as not turn out to be deep-
seated political, economic, or cultural, problems and obstacles that
are notoriously difficult to overcome.

These factors have stubbornly refused to yield to the best-
conceived and most well-intentioned efforts in many fields. They
have long prevented a more rational and equitable balance between
population and resources, and it seems inevitable that they will do so
in the future as well. Indeed, one sometimes wonders if the present
arrangements are not the least rational imaginable! If, for example,
today's system of international trading in goods and services was
more tolerant of the products of the developing countries, most
development aid might be unnecessary. However, this is not at all
today's reality.

Thus, even if it were true that, from a global point of view, the
Earth could sustain many times more people than it does today, this
knowledge would be of little help because the measures taken, and
policies pursued, to satisfy human needs are rarely global in scope
and intent. Moreover, for reasons discussed above, even when the
policies are global, they are all too often ineffective or unsuccessful.
The reality is that we are confronted not by a single population
growth-problem but by many, including scores of national, regional,
and even continental, population growth-problems. When countries
experience serious imbalances between population and resources, a
reduction in population through migration is, generally speaking, not
an available option. No way has been found of gaining general
acceptance for the peaceful transfer of 'surplus' populations from
one country, region, or continent, to another. The other major option,
the transfer and importation of resources to correct the imbalance,
has had success in some instances. However, this is neither an easy
nor a permanent solution, as is shown by the persistence of hunger,
undernourishment, and poverty, in so many countries today.

In the area of population programmes, however, the answers are
very clear. We have the excellent guidance contained in the ICPD
Programme of Action (1994). The most important point is that, no
matter what our policy-goals may be in regard to population, the
only way to reach them is by creating the conditions in which men
and women can effectively make and carry out their own decisions
related to reproductive health and family planning. I will return to
these points briefly at the end of this chapter.

POPULATION GROWTH, RESOURCE SCARCITY, AND SOCIAL CHANGE

Under conditions of population growth, a given amount of resources will give rise to increased competition for those resources. The intensity of the competition will depend upon the relative scarcity and the need or usefulness of the resources in question.

In the past, lower population numbers and, consequently, less exploitation of resources, allowed renewable resources to regenerate. Now, however, the exploitation of many resources, fuelled by vastly higher and rapidly increasing population numbers, and by gross inequities of economic power, goes beyond regenerative capacities. In many cases, the rate of exploitation is well above the threshold at which any further harvesting will damage the resource. More people simply require more food, fuel, and clothing, which must come from the Earth's resources (Tolba *et al.*, 1992 Ch. 16).

Of all resources, land is in a category of its own. Serving as the source of most of the food we eat, as the site for the dwellings we erect, and as a repository of most other resources we depend on, land also has a value in and of itself, independently of the resources that it may contain and the economic calculations to which it may give rise.

According to a recent FAO report:

'The competition for *land* is projected to intensify between sectors and production systems. It is expressed most accurately in the expansion of the use of land for arable and tree crops, shifting cultivation, grazing livestock, and its conservation under forest. Then there is the competition between crop and livestock production and, on a much smaller scale, between crop production or mangrove swamp preservation and aquaculture: and... there will be further pressures on the forest for timber and fuel-wood extraction. Finally, increasing population and economic growth will contribute to further diversion of land to human settlements and infrastructure' (FAO, 1993 pp. 271–2).

This competition will become no less strong as the phenomena of land degradation and desertification proceed, to which population pressures also contribute. 'Desertification (broadly: land degradation in dryland areas) is estimated to affect some 30% of the world's land surface' (FAO, 1993 p. 16). According to the United Nations Environment Programme (UNEP), '[by] the end of this century, it is estimated that the world will have lost a substantial portion of its arable land if the current advance of desertification is not arrested' (UNEP, 1992 p. vi).

Of all the uses to which land is put, the production of food for human consumption is claiming an increasing share of the available terrain. As land degradation spreads and population increases in many developing countries, the pressure builds up to convert to agricultural use and human settlements more and more land that, so far, has not been used for these purposes. In areas of the world that are already suffering from severe land-scarcity, this conversion of land is likely to become a critical issue. It is estimated that in South Asia, for example, 40% of still-unused land with agricultural potential may have to be converted to agricultural use and human settlements over the next couple of decades. After the year 2010, there will be little room for further expansion, at least for human settlements (FAO, 1993 p. 15). Yet, the population of the area is projected to be still growing at a rate of 1.5% a year in AD 2010, at 1.3% in AD 2015, and at 1.1% in AD 2020.

So far, the growth of world food production has kept ahead of population growth. During the 1970s and 1980s, food production grew at an average annual rate of 2.3%, whereas population growth averaged about 1.8–1.9%. Over the next decade-and-a-half, until AD 2010, food production is estimated to grow by perhaps 1.8% a year, while world population growth is expected to decline from an annual average of 1.7% in the 1990s to an annual average of 1.4% during the first decade of the next century. The most salient point is that the food production growth-rate is declining faster than the population growth-rate. Some parts of the world, such as sub-Saharan Africa, are now 'worse off nutritionally than 30 or 20 years ago' (FAO, 1993 p. 2).

The pressures of population growth and the demands particularly from industrialized countries, are also threatening the vital role of forests as depositories of scarce ecosystems and biodiversity. Intact forest ecosystems and ecocomplexes perform a variety of important functions and economic services, including acting as gene-pools, water regulators, and flood controls, as watersheds (preventing soil erosion), as protectors of inland and coastal fisheries, and as climate stabilizers. They also serve as recreation areas (Durning, 1994 p. 33). The growth of population generates not only additional demand for forest products but also 'increasing pressures for transfer of forest land to agriculture, infrastructure, and urban uses' (FAO, 1993 p. 153).

Earlier fears of overexploitation of non-renewable resources have been laid to rest in some cases by the emergence of substitutes. However, because the 'rules' applying to non-renewable resources are wholly different from those governing renewable resources, the fears now surrounding the latter are indeed warranted. 'It is an odd turn of language by which those resources called "nonrenewable" are the ones we will *not* run short of, while "renewable" resources like forests and water will be the ones whose shortage limits us' (Keyfitz, 1991 p. 42). If, after having been encroached upon, they are to return to anything near their previous state, tropical ecosystems, for example, can probably tolerate relatively modest inter-ference and exploitation. However, exploitation of tropical forests in many parts of the world is so extensive and intrusive as to dispel all hope that, in the areas concerned, this resource can be renewed in any true sense of the word. When 'renewal' is at all attempted, it is at the cost of ever-decreasing diversity.

The population–resource balance is not solely a land-based phenomenon. The living resources of the seas and oceans have long been under tremendous pressure, and over-fishing is a world-wide problem. In just three years, from 1989 to 1992, the world's total fish-catch declined from 86 million tons to 80 million tons (Weber, 1993 p. 32). Given that the living resources of the seas and oceans are finite, and given also that we have reached (or inevitably will reach, if the right policy-changes are introduced) the maximum sustainable yield, no degree of redistribution and application of prin-ciples of equity and fairness can hide the fact that population growth will inevitably lead to a fall in the supply of edible fish per person.

The excessive exploitation of the seas' resources occurs often in areas with the highest population density — near the coast. More than half of the world's human population is estimated to live within 100 kilometres of a sea-coast. Areas with especially high coastal population concentrations include South-East Asia — where more than two-thirds of the people live in the coastal zone — China, South Asia, Europe, southeastern Africa, and parts of North and South America. Coastal populations appear, in general, to grow more rapidly in numbers than do non-coastal populations. This is due in no small measure to rapidly-growing urbanization, especially in the developing world. Of the world's 10 largest cities, 9 are coastal, and of the 50 largest cities two-thirds are coastal. Nearly 20% of the world's population lives in coastal cities. This means that a growing portion of the world's population is exploiting the resources of the sea for which the 'potential for increasing total production much above present levels is rather limited' (FAO, 1993 p. 175).

MAJOR SHIFTS OF POPULATION

Today, there is no lack of signs and indicators that speak for a some-times drastic shift in the balance between population and resources, generally to the detriment of the resources and often drastically so. But rapid population-growth has other effects that are becoming increasingly visible. As mentioned earlier, land has a value in and of itself which goes beyond the resources it may contain and the economic calculations to which it may give rise. The possession of land or territory is a defining characteristic of a state, and the tenacity with which states cling to territories that are proven liabi-

lities in economic and other terms is just one indication of the
symbolic and emotional value of land. (Russia's sale of Alaska to
the United States in 1867 was a rare exception.)

In centuries past, the possession of land and territory was a
foremost criterion of power and influence and, therefore, a frequent
cause of campaigns of war and conquest. Today, in contrast,
economic prowess and technological sophistication count for more
in this respect. Land, however, still retains its importance. In the
past, the population problem in many parts of the world was to find
enough people to settle vast, empty spaces. Today's problem is to
find space for hugely increased and rapidly growing populations.

Peoples' attachment to land includes strong proprietary sentiments. These are sometimes given expression in the idea of 'the sons
of the soil', who are seen to have a prior right to the disposal and use
of the land because they were the earliest occupants. In some
countries this idea has assumed the status almost of national ideo-
logy, and it is more or less present, in one form or another, in almost
all countries. It is reflected, for example, in the land-rights granted in
recent years to indigenous peoples in the settler states of North
America and Australasia. These rights may be seen as a form of
restitution for 'sons of the soil' rights, of which the indigenous
peoples were deprived by the European settlers and colonizers in
centuries past.

The influx, for whatever reason, of new people into an area
traditionally settled by a particular ethnic or national group, is often
seen as a threat to historical and traditional proprietary rights and
sentiments. Many times, population growth and intolerable density
in one part of a country have caused people to move to another part,
only to be received by the local population with resentment and
hostility that may also stem from differences in ethnic allegiances.
Even stronger reactions are usually evoked when people move from
one country to another.

For these and other reasons there are indications that migration
and immigration are increasingly seen internationally in negative
terms, and even as threats to national security and stability. According to one report, in 1976 only 6.4% of all nations considered their
immigration levels to be too high. By 1989, this figure had risen to
20.6% and 'as many as 31.8% of states hinted that they would like
to put a brake on immigration so as to lessen pressures on national
borders' (Widgren, 1990 p. 749). In 1992, the World Bank estimated
the number of international migrants of all kinds to be about 100
millions, and the number of internal migrants, whether rural–urban,
urban–rural, or rural–rural, to be far higher (UNFPA, 1993 p. 7):

'Continuing rapid population growth in many parts of the developing
world; high levels of natural increase in cities as well as continuing

rural–urban migration; the addition of unprecedented numbers of young people, many with some education, to the urban labour-force; continuous contact with the values and lifestyles of more affluent countries, coupled with a general rise in expectations, indicate the likelihood of more rather than less international migration in the future' (UNFPA, 1993 p. 15).

Whether internal or international, migration is a symptom of social transformation. Perhaps the most striking characteristic of social transformation processes today is the speed with which they proceed compared with the more leisurely pace of times past. Coping with rapid change, however, is not easy. Technology, for example, both creates and destroys employment opportunities.* Even without large added increments of people entering the work-force every year, the industrialized countries have found it exceedingly difficult to encourage or provide new employment for people whose parts in the work-process have been taken over by the introduction of new technology. In many parts of the developing world, however, technology is less important as a contributor to unemployment than is population growth, which every year adds many more people to the work-force than the economies are able to absorb. It was no accident, therefore, that the creation of employment opportunities was one of three major themes addressed by the World Summit for Social Development, which was held in 1995 in Copenhagen, Denmark.*

Of all the social transformation processes which are taking place today, rapid urbanization is the most consequential. The world's urban population is expected to reach 3.2 thousand millions by the year 2000 and 5.5 thousand millions by the year 2025 (out of a total of 8.5 thousand millions). In the year 2000, the developing-countries' share of the urban population is projected to be 70% and, by AD 2025, it is expected to reach 80% (UNDP, 1991 pp. 10 & 12). It is generally recognized that no factor — not even rural–urban migration — contributes more to the growth of urban areas than the natural increase of the already urbanized population.

Urbanization may be described as the process whereby sufficient facilities, services, and employment opportunities, are provided to a growing number of people within an urban setting. Seen in this light, many parts of the world — the developing world, in particular — are experiencing, not real urbanization as described above, but a

---

* *E.g.* with machines replacing people ever more widely. It is no accident, therefore, that the creation of employment opportunities was one of the major themes addressed by the Programme of Action of the World Summit for Social Development held in 1995 in Copenhagen, Denmark (*see* Chapter 3 of the Programme of Action: *Expansion of Productive Employment and Reduction of Unemployment*).

relentless concentration of people in a more or less confined physical space in which the barest of amenities are often lacking. The
reason is simple. The growth of human numbers in urban settings is
far outstripping the capabilities of Governments to provide infrastructure, services, and employment.

PROSPECTS FOR GLOBAL STABILITY

As mentioned earlier, stability has not been, and is not, a globally
pervasive condition. This is just another way of saying that we have
not yet experienced such a thing as global stability. The domain over
which the condition of stability has prevailed, has always been partial and incomplete and never global in scope. Now, as always, the
great political task is to extend the geographical domain over which
a desirable degree of stability prevails.

Social and environmental change of the sort described here is
taking place on a scale that has never been witnessed before, and it
may both add to and detract from such stability as exists today.
Experience tells us, however, that change of this magnitude, involving so many and growing numbers of people, is unlikely to take
place without social tension and painful adjustments. It embraces the
concerns of thousands of millions of people, and no Government or
society is untouched by it. To cope with these changes, Governments
need resources and capabilities which, in all too many cases, fall
seriously short of what are available or are simply completely
lacking. If support for the most disadvantaged developing countries
(and there are many in or near that position) is not forthcoming in
the years ahead, it seems likely that instability and disorder will be
experienced on a much larger scale than they are even today. The
adoption in 1992 by the United Nations Conference on Environment
and Development (UNCED) of the Rio Declaration and *Agenda 21*
was, *inter alia*, an implied recognition of this possibility.

In the population field, much changed in the time between when
the last large-scale United Nations population conference was held in
Mexico City in 1984 and the International Conference on Population
and Development (ICPD) was held in Cairo in September 1994.
First, there had been a large influx of new members of the United
Nations since 1984, when the membership stood at 159, as against a
membership in 1994 of 184. The demographic conditions of many of
these new states were and still are precarious. The ICPD offered
these states the opportunity to participate in an international consen-
sus-building endeavour on issues of great importance to their future.

Second, many other states have faced changed, and in several
cases worsened, demographic circumstances, and have redefined

their attitudes towards the population issue. These new attitudes needed to be integrated into a new consensus.

Third, after 1984, the development paradigm, responding in particular to the World Commission on Environment and Development and UNCED, underwent changes that inevitably had an impact on the population field as well.

Lastly fourth, by the time of the ICPD, it was widely accepted that rapid population growth exacerbates rather than relieves the environmental and social problems which the world is facing so widely today.

The ICPD was extremely important in terms of galvanizing this awareness and drawing from it the appropriate conclusions. The Conference agreed on a landmark Programme of Action for the next 20 years in the field of population and development. Recognizing the crucial contribution that early stabilization of the world population would make towards the achievement of environmentally sustainable development, the Programme also recognizes that stabilization can be achieved only by taking individual people's perspectives into account, and by ensuring the full and, wherever appropriate, equal participation of women in all aspects of development. It moves away from a focus on fertility towards a comprehensive approach that integrates family planning with reproductive health, and addresses a wider range of concerns — particularly economic status, education, and gender equity and equality.

The ICPD Programme of Action reaffirms the basic human right of all couples to decide freely and responsibly the number and spacing of their children, and to have the information, the education, and the means, to do so. It emphasizes that men have a key role to play in bringing about gender equity and equality, in fostering women's participation in development, and in improving women's reproductive health. Goals are set out in three related areas: expanded access to education, particularly for girls; reduced infant, child, and maternal, mortality rates; and increased access to quality reproductive health services, including family planning.

Many of the recommendations are aimed at strengthening and supporting families. The Programme recognizes that both married and unmarried adolescents need sex education and counselling services to protect them from unwanted pregnancies and sexually-transmitted diseases, and that satisfaction of these needs should involve parents.* Special attention is paid to internally-displaced persons and international migrants, among others, and to easing the

---

* Here as widely elsewhere we look in vain for any reference to self-discipline and personal restraint in matters of sexual indulgence such as older generations were brought up to — so that teen-age pregnancies and single-parent families were virtually unthinkable in educated circles in our youth. — Ed.

pressure that rapid urbanization puts on social infrastructure and local environments.

The ICPD Programme of Action urges all countries to make reproductive health-care and family planning accessible, through their primary health-care system, to all individuals of appropriate ages no later than AD 2015. The financial resources required for this effort in the less-developed countries and those with economies in transition, is estimated at $17 thousand millions for the year 2000, increasing to $21.7 thousand millions by AD 2015. Developing countries will continue to meet up to two-thirds of these costs themselves; approximately one-third — $5.7 thousand millions in AD 2000 and $7.2 thousand millions in AD 2015 — will have to come from external sources (United Nations, 1994).

Although ambitious, the ICPD Programme of Action by itself is insufficient to ensure the degree of international stability on which progress towards environmentally sustainable development is vitally dependent. Other political, economic, and environmental, measures would also have to be brought into action — such as those contained in *Agenda 21*, those that emerged from the World Summit for Social Development, the Fourth World Conference on Women, and the Habitat II Conference in June 1996. The issues involved in these and other conferences are intimately linked, and it should be a priority task to weave them all into a coherent framework for international action. In this picture, due consideration of population factors may not alone be sufficient to achieve broader development goals, but such consideration will most certainly be necessary.

REFERENCES

DURNING, A.T. (1994). Redesigning the forest economy. Pp. 22 – 40 in *State of the World 1994* (Eds L.R. BROWN *et al.*). W.W. Norton & Co., New York, NY, USA (for the Worldwatch Institute): xvi + 266 pp., illustr.

FAO (1993). *Agriculture: Towards 2010*. Food and Agriculture Organization of the United Nations, Rome, Italy: viii + 320 pp. + Appendix 42 pp.

ICPD PROGRAMME OF ACTION (1994). *See* UN GENERAL ASSEMBLY (1994).

KEYFITZ, N. (1991). Population growth can prevent the development that would slow population growth. Pp. 39–77 in *Preserving the Global Environment* (Ed. J.T. MATHEWS). W.W. Norton & Co., New York, NY, USA (for the American Assembly and World Resources Institute): 362 pp., illustr.

MATHEWS, J.T. (1991). Introduction and overview. Pp. 15–38 in *Preserving the Global Environment* (Ed. J.T. MATHEWS). W.W. Norton & Co., New York, NY, USA (for the American Assembly and World Resources Institute): 362 pp., illustr.

SENANAYAKE, P. (1995). Guest Comment: Reflections on the UN's International Conference on Population and Development. *Environmental Conservation*, 22(1), pp. 4–5.

TOLBA, MOSTAFA K., EL-KHOLY, OSMA A., EL-HINNAWI, E., HOLDGATE, M.W., McMICHAEL, D.F. & MUNN, R.E. (Eds) (1992). *The World Environment 1972–1992: Two Decades of Challenge*. Chapman & Hall, London, England, UK (for United Nations Environment Programme): xi + 884 pp., illustr. and tables.

UNDP (1991). *Cities, People and Poverty: Urban Development Cooperation for the 1990s.* United Nations Development Programme, New York, NY, USA: vi + 94 pp., illustr.

UNEP (1992). *World Atlas of Desertification.* Edward Arnold, New York, NY, USA (for United Nations Environment Programme): x + 70 pp., illustr.

UNFPA (1993). *The State of World Population 1993: The Individual and the World: Population, Migration and Development in the 1990s.* United Nations Population Fund, New York, NY, USA: ii + 54 pp., illustr.

UN GENERAL ASSEMBLY (1994). *Draft Final Document of the Conference: Draft Programme of Action of the Conference: Note by the Secretary General.* (A/CONF. 171/PC/5, 18 February 1994.) Preparatory Committee for the International Conference on Population and Development, New York, NY, USA: 82 pp.

UNITED NATIONS (1995). *Report on the International Conference on Population and Development (Cairo, 5–13 September 1994)* (A/CONF. 17113). United Nations, New York, NY, USA.

WEBER, PETER (1993). *Abandoned Seas: Reversing the Decline of the Oceans.* (Worldwatch Paper 116.) Worldwatch Institute, Washington, DC, USA: 68 pp., illustr.

WIDGREN, JONAS (1990). International migration and regional stability. *International Affairs*, **66**(4), pp. 749–66.

WORLD SUMMIT FOR SOCIAL DEVELOPMENT (1995). (Held in Copenhagen, Denmark.) Programme of Action Chapter 3: *Expansion of Productive Employment and Reduction of Unemployment.* [Not available for checking.]

2

Global Population
and Emergent Pressures

by

NORMAN MYERS, MA (Oxon.), PhD (UCB)

Visiting Fellow, Green College, Oxford University,
and *Senior Fellow, World Wildlife Fund-US;*
Consultant in Environment and Development,
Upper Meadow, Old Road, Headington, Oxford OX3 8SZ,
England, UK.

INTRODUCTION

Practically all students of the world scene have latterly come to agree that humanity now faces severe emergent problems owing to the continued rapid growth in human numbers and their takings from Earth's limited 'cake' (Union of Concerned Scientists, 1992; US National Academy of Sciences and [British] Royal Society, 1992). Yet even apart from biassed communities and other fanatics, there is still considerable debate about the most desirable, and practical, relationship between population, environment, and sustainable development*, but commonly means favourable-for-humans development that is sustained at least to the extent of not seriously harming the local environment. Much light was thrown on these topics at the United Nations' Conference on Population and Development, which was held in Cairo, Egypt, in mid-September 1994 (*cf.* Senanayake, 1995).

The present chapter reviews some of the more salient analyses, findings, and conclusions, concerning the multiple emergent pressures stemming from population — whether from the present population total or from its egregious growth-rate especially in certain Third-World countries.

The chapter accepts that the relationships between population and environment are multifaceted and complex — nowhere near as straightforward as has sometimes been represented. Many other variables are at work, such as negligent technologies, defective markets, inefficient economies, and faulty policies overall. But the

---

* Evidently meaning, in this context, development that is favourable for humans and is sustainable without seriously harming the local or global environment. — Ed.

chapter firmly postulates that population is a prominent factor —
often a predominant factor — in many emergent problems of
environmental decline and unsustainable development.

By way of illustration, consider the prospect of sub-Saharan
Africa. The mid-1996 population of 597 millions is projected (not
predicted, still less forecast) to grow to 1.3 thousand millions by AD
2025, and to quadruple or even quintuple by the time it attains
stationary numbers roughly 150 years hence (World Bank, 1991;
United Nations Population Fund, 1992a). Already one person in
three of its inhabitants is malnourished. Yet the region's current
food-deficit of 12 million tonnes per year is expected to rise to 50
million tonnes (one-third of anticipated consumption) by the year
2000 and to 250 million tonnes by AD 2020 (roughly equal to all of
the Maize [*Zea mays*] grown annually in North America) (Braun &
Paulino, 1990; Pinstrup-Andersen, 1992; *see also* McNamara, 1990;
Schreiber & Cleaver, 1992). To put these figures in perspective,
recall that food aid world-wide today is only 7.5 million tonnes.

Sub-Saharan Africa's hopes of purchasing food from outside are
meagre in the light of its trade relations. Commodities comprise 90%
of the region's exports, and while the volume of these exports
increased by 25% during the 1980s, the revenues declined by 30%
owing to slumping prices on world markets. In addition, the region
pays out the equivalent of more than US $1 thousand millions every
month (twice as much as it receives in aid) to service a debt-burden
that proportionally is three to four times as heavy as Latin America's.

Were the region enabled, however, to double its annual increase
in food production, and to cut its annual population growth-rate in
half by AD 2020, it would then become food self-sufficient
(Pinstrup-Andersen, 1992). In other words, although the region's
problems are dire, solutions are available. The resources in shortest
supply are human innovation, appropriate technology, external
funding among other supports, policy responses backed by political
will, and, above all, time to mobilize these diverse resources.

BENEFITS OF REDUCED POPULATION-GROWTH

The benefits of reduced population-growth are so profound and pervasive that it is worth-while to review them before going on to the
chapter's main analysis. The impact of population growth can be
represented in its essentials through an equation I = PAT, where I
stands for impact, P for population, A for affluence (*per caput* level
of material living and consumption), and T for technology to sustain
that affluence — especially technology of any environmentally
adverse sort. Note that the three factors P, A, and T, interact in

multiplicative fashion, *i.e.* they compound each other's impacts (Ehrlich *et al.*, 1989; Ehrlich & Ehrlich, 1990).

The above equation immediately makes clear why there are population problems not only in developing countries but also in developed countries where the A and T multipliers for each person are unusually large. The equation also makes clear why developing nations with large populations, albeit with little economic advancement, can generate disproportionate impact on the planetary eco-complex* by virtue of the fact that the P multiplier on the A and T factors can be exceptionally large (Ehrlich *et al.*, 1989).

Consider, for instance, the large-scale repercussions that stem from small *per caput* amounts of coal-burning or chlorofluorocarbon (CFC) use in China and India with their collective total of more than two thousand million people today. The large populations make it difficult to control overall emissions of CO_2 and CFCs during the course of the two countries' industrial expansion — even though global warming and stratospheric ozone-shield depletion may well impose marked injury on the agriculture of these two nations (as well as similar injury elsewhere), with their populations projected soon to comprise almost two-fifths of Humankind. Consequently, there can be no realistic strategy to reduce their emissions of 'greenhouse' gases and CFCs without, among other measures, population planning of a scope and with an urgency far greater than has been exhibited to date.

Indeed practically all countries, whether undeveloped, developing, or developed, are experiencing major difficulties in sustaining their populations on their own environmental resources — particularly in regard to soil stocks, water supplies, forests and grasslands, fisheries, the atmosphere with its limited pollutant-absorptive capacity, and climate. Because of environmental degradation, countries of Western Europe and North America are losing around 4% of their Gross Domestic Product (GDP), countries of Eastern Europe and the former Soviet Union are losing between 6% and 10%, and many developing countries are losing between 12% and 18% (Pearce & Atkinson, 1992).

Much of this environmental degradation is due to population growth in conjunction with resource-wasteful life-styles and negligent technologies (but note that, while many developing countries have inadequate population policies, not a single developed country has any population policy at all, if we disregard those few countries that still see fit to adopt a pro-fertility stance). The population linkage applies whether populations are growing quickly or slowly,

---

* Not 'ecosystem' as some are wont to say — *cf.* 'On the Use and Misuse of the Term "Ecosystem",' by N. Polunin & E.B. Worthington, *Environmental Conservation*, **17**(3), p. 274, 1990. — Ed.

though the countries with high population growth-rates (1.5% per year or more) are almost invariably among the developing or less-developed countries. How far they experience environmental degradation because they are poor, and how far they are both poor and suffer environmental degradation because they have high population growth-rates, is the subject of much discussion. There is abundant evidence in support of both contentions; and both are intimately interconnected through a host of feedback processes.

Enough Food for All?

To illustrate further the resource/population connection, consider the most basic of human needs, namely food. During the period 1950–84, grain production grew by an average of 2.9% per year while population growth averaged around 2.0% per year. But from 1985 to 1992, there was far less annual increase in grain production, even though the period witnessed the world's farmers investing thousands of millions of dollars to increase their output. Crop yields 'plateaued'; for plant breeders and agronomists had (temporarily?) exhausted the scope for technological innovation. The 1996 harvest was only 9.1% higher than that of 1985, yet there were an extra 900 million people to feed! While world population increased by 19%, grain output *per person* declined by 9% (Brown *et al.*, 1992, 1993, 1997).

As for the future, note that if ever there are as many as 10 thousand million people to be fed adequately (the United Nations *medium projection* is of that number in AD 2050), we shall have to produce nearly three times as many calories as we do today. To grow that much food, we shall need to farm all the world's current croplands as productively as Iowa's best cornfields, or three times the present world average (Repetto, 1987; *see also* Kendall & Pimentel, 1993). Regrettably, there are all-too-few new areas to be opened up for agriculture,* while all-too-many are being lost to highways and other 'development'. *Per caput* cropland expanded at an average annual rate of 0.5% per year throughout the period 1950–80, but since then the rate has been only half as much; and primarily because of population growth, the area *per caput* of arable land has *declined* by an average of 1.9% per year. Similarly, irrigated lands — which supply one-third of our food from one-sixth of our croplands — grew by 2–4% per year during 1950–80, but have grown by an average of only 1% per year since 1981 (Brown *et al.*, 1992, 1997).

---

* Hearing on the morning of 3 December 1996 the BBC's World News broadcast about an indication of possible ice — and hence water — on the Moon, and there being seemingly no limit to Humankind's ingenuity if not much else, one's imagination surges to the conceivability of Earth-relieving agriculture or at least pisciculture on what the *Encyclopaedia Britannica* refers to as 'the natural satellite of the planet Earth' or elsewhere in 'outer space'. — Ed.

Water Deficits

Further resource/population constraints are emerging, this time with regard to water. Demand for fresh water is expected to double between AD 1971 and 2000 in nearly half the world's countries, simply to cater for the needs of additional people (Falkenmark & Suprapto, 1992). Already five hundred million people experience chronic water-shortages, and the total is projected to reach three thousand millions by AD 2025 through population growth alone (*ibid.*). Water shortages exacerbate problems of public health as well as agriculture. A full 90% of disease in developing countries is associated with lack of clean water for domestic use, and the chief sufferers are children. As we shall see below, child mortality is a prime factor in the reluctance of many developing-country women to engage in family planning.

Tropical Deforestation

Similarly, population growth is considered to have been the chief factor in tropical deforestation during the past several decades (Myers, 1991*a*). Further, and according to an assessment from the Food and Agriculture Organization (1993*a*), population growth will result in the annual elimination of at least 100,000 square kilometres of forest for agricultural expansion alone, even allowing for the most optimistic assumptions of increasing food production and agricultural efficiency. Thus population growth will be accounting for almost three-fifths of tropical deforestation during the foreseeable future (Myers, 1993*a*; *see also* Grainger, 1993).

Unemployment

Furthermore, population growth causes not only environmental problems but problems of economic and social kinds that also relate to development, *e.g.* unemployment. Today the developing countries' work-force numbers two thousand million people. Of these, at least one-third are unemployed or grossly underemployed; their total exceeds the entire work-force of the developed countries. By the year 2025, and because of population growth, the developing countries' work-force is projected to expand to well over three thousand millions (World Bank, 1990). To supply employment for the new workers, let alone those without work today, the developing countries will need to create almost 40 million new jobs each year during the 1990s (Purcell, 1993). The United States, with an economy half as large again as the entire developing world's, often has difficulty in generating an additional two million jobs each year. Many of the developing world's unemployed find they can gain a livelihood only by cultivating marginal lands such as forests, semi-arid savannas, and hilly terrain, thus causing deforestation, desertification, and soil erosion (Little & Horowitz, 1987; Leonard *et al.*, 1989; Kates & Haarman, 1991; Mink, 1991).

Of course, we must be careful not to oversimplify the situation. Several other factors are involved in the nexus of population-related issues, most notably negligent technology, meagre infrastructure and public services, market deficiencies, inadequate savings and investment, and faulty policies generally. It is difficult to separate out the specific and precise role of population growth in problems of environment and development: the 'linkages calculus' is attended by uncertainty of many sorts. But there is much evidence (Repetto, 1987; Ehrlich & Holdren, 1988; Shaw, 1989, 1992; Ehrlich & Ehrlich, 1990, 1991; Holdren, 1990, 1991; Daly, 1991; Davis & Bernstam, 1991; Keyfitz, 1991*a*, 1991*b*, 1992; Myers, 1991*b*, 1992*a*, 1993*a*, 1993*b*; Dasgupta, 1992; Kessler, 1992; Tolba *et al.*, 1992; United Nations Population Fund, 1992*b*, 1992*c*; Ness *et al.*, 1993; Polunin & Burnett, 1993) that population growth is engaged with other issues through multiple relationships, wherein one problem serves to compound the adverse impacts of others — and whereby a solution to one problem can reinforce solutions to other problems.

Notwithstanding this overall conclusion, the factor of uncertainty is so central to the overall analysis that it is worth brief elaboration, especially with regard to policy responses. What is 'legitimate scientific caution' in the face of manifold uncertainty, especially in so far as *uncertainty can cut both ways*? Some scientists may assert that, in the absence of conclusive evidence, it is better to keep to a cautious evaluation of population linkages on the grounds that it is 'more responsible'. But note the crucial factor of asymmetry of evaluation. A cautious evaluation, ostensibly 'safe' because it takes a conservative view of such limited evidence as is at hand in documented detail, may fail to reflect the real situation just as much as does an 'unduly expansive' evaluation that is more of a best-judgement response based on all available evidence with varying degrees of demonstrable validity. A minimalist evaluation with apparently greater precision may amount to spurious accuracy. In a situation of uncertainty where not all linkages can be demonstrated to conventional satisfaction, we should not become preoccupied with what can be accurately counted if that is to the detriment of what ultimately counts. Undue caution can amount to recklessness; and as in other situations beset with uncertainty, it will be better for us to find that we have been roughly right than precisely wrong!

POPULATION, ENVIRONMENT, AND DEVELOPMENT: INTERACTIVE RELATIONSHIPS

Size of Global Population and Its Growth-rate

The global population in late 1996 amounted to 5.8* thousand million people. Of these, 4.6 thousand millions were in un-developed or 'developing' nations and 1.2 thousand millions were in 'developed' nations (United Nations Population Fund, 1993; World Bank, 1993). The total was growing at a rate of 1.5% per year. Whereas the growth-rate declined from 2.1% in 1970 to 1.8% during the seven years to 1977, it has declined during the past 16 years by only 0.1% to its present level of 1.7%.

A global annual population growth-rate of 1.7% exerts a formid-able impact world-wide. Suppose that *per caput* resource con-sumption were to be reduced by 5%, and improved technologies were to cause 5% less environmental damage, but these two ad-vances were to occur in a world where the population growth remained at 1.7% per annum, then the reductions in environmental impact would be cancelled out in less than six years (Ehrlich & Ehrlich, 1990).

What is also critical to the population prospect is the absolute increase in numbers. The current annual increase of 90 million people is projected to persist until at least the year 2025 (United Nations Population Fund, 1992a). Of the annual increase, 95% is in developing nations, which have only limited capacity to cope with the environmental and economic consequences of rapid growth in human numbers — owing to their unproductive agriculture, low *per caput* incomes, low-level technology, and inadequate savings and investment. Even if they enjoyed all these requisites, they would be hard-pressed. A country with the planning capacity of Switzerland could hardly accept twice as many people in a single generation, let alone do it time and time again. Emergent pressures indeed!

Future Population Growth: Key Determinants

So far as we can discern, fertility** is determined by a number of factors, including socio-economic advancement†, women's status, child mortality, and family planning. The importance of each factor varies from society to society, as we shall see below.

---

* Seemingly most authoritatively updated in Sadik (1996). — Ed.

** In the sense of the number of children produced by the average woman of repro-ductive age. — Ed.

† According to a practising psychiatrist of our acquaintance, another factor could be the belief of many modern women that they will be considered social failures if they do not bear at least two children. — Ed.

Consider the first-mentioned factor of *socio-economic advancement*. There seems to be a strong association, if not correlation, between rising prosperity and falling fertility-rates. But not always: Sri Lanka's fertility rate is 2.3 and its *per caput* GNP is US $640, whereas Brazil, with a *per caput* GNP of US $3,370, has a fertility rate of 2.8. So an additional critical factor appears to be seated in social equity. In Sri Lanka the income ratio of the bottom 20% *vis-à-vis* the top 20% of the population is 1:9, but in Brazil it is 1:28 (World Resources Institute, 1992). A situation similar to Sri Lanka's applies to China, Java in Indonesia, Kerala State in India, Cuba, and several other societies that are characterized by an unusual degree of egalitarianism.

Specially pertinent is the position of the most impoverished sector of Humankind, namely the 1.2 thousand million people with a cash income of less than the equivalent of US $1 per day. They feature the highest fertility-rates, and their total number is expected to keep on increasing until the year 2000, or possibly much longer, unless there is a far more substantive attack than at present on their absolute poverty. Not much better off are the 1.7 thousand million people with a cash income of from US $350 to $500 per year; and they, too, tend to feature high fertility-rates. These two categories, constituting more than half of Humankind, account for well under 5% of global income (Haq, 1992; Myers, 1994). But note that Bangladesh, one of the poorest countries on Earth, doubled its contraceptive use during the 1980s (Abernethy, 1993), offering further reason to believe that sheer socio-economic advance is not an invariable prerequisite for fertility decline.

Next, consider *women's status*. There are twice as many illiterate women as men in developing countries, their health is usually much worse, their employment outlook is poor at best, they own less property than men, and their meagre inheritance-rights place them still more at a disadvantage (World Bank, 1990; United Nations Development Programme, 1993). There is much evidence (Grant, 1992, 1993; United Nations Population Fund, 1992*a*, 1993) that, when women receive enhanced status in these respects, *plus* increased social standing all around, they readily engage in reduced fertility. Despite some brighter prospects indicated in the preceding chapter, however, there is little sign that women, especially in many Third-World countries, can look for much improvement in their lot in the foreseeable future — unless, of course, there is an explicit effort to tackle the problem on grounds of basic human rights*, in

---

*As was courageously advocated by Dr Nafis Sadik at the recent World Conference on Human Rights, held in Vienna, Austria. Dr Sadik (contributor of the preceding, opening chapter in this anthology, is Executive Director of the United Nations Population Fund (UNFPA) and was Secretary-General of ICPD. — Ed.

which case there could be a significant shift in this key determinant of future population growth.

Third, consider *child mortality*. The annual death-toll of children is some 14 millions per annum, and is declining by only about half-a-million per year (Grant, 1993). As long as parents find that a substantial proportion of their children die, they remain less inclined to restrict their fertility (Grant, 1992; United Nations Population Fund, 1992a; World Bank, 1993).

Finally, consider *family planning*. Of roughly 750 million women 'at risk' in developing countries in 1990, almost half were not using any form of modern contraception. During the current decade, their total number is projected to increase by more than 200 millions. To achieve an average family-size of 2.8 children (still a lot higher than replacement fertility), the number of contraceptive users will have to increase by 61% (Sinding, 1992; United Nations Population Fund, 1992a). But external 'population assistance' declined from 1.3% of Official Development Assistance (ODA) in 1986 (a slight-enough level, given the handsome benefits of higher investment!) to 0.9% in 1991 (Sinding, 1992).

Because of deficiencies on these various fronts, many of the so-called developing countries are failing to make progress to a demographic transition. Rather are they caught in a 'demographic trap' wherein it is precisely their population growth that denies them the economic growth which would enable them to reduce their population growth (Ehrlich & Ehrlich, 1990; Keyfitz, 1991a). A poor country with a population growth-rate of 2.5% per annum needs an economic growth-rate of 5.0% in order to achieve the 2.5% *per caput* economic growth that is often regarded as the minimum for 'passage through the demographic transition'. While an *economic* growth-rate of 5.0% has been achieved by a number of countries, generally accompanied by declining fertility, many other countries have managed little economic growth since 1980 (World Bank, 1992). In the meantime their persistently high population growth-rates consolidate their poverty — fuelling further emergent pressures all around.

Environment: The Connection to Population Growth

Population pressures are expressed through environmental decline of multiple sorts. In recent single past years alone we on Planet Earth have:

- lost 25 thousand million tonnes of topsoil, or enough in principle to grow nine million tonnes of grain and thus to make up the diets of at least 200 million undernourished people (Brown *et al.*, 1993; Pimentel, 1993).

- lost 150,000 square kilometres of tropical moist forest, with significant costs in terms of depleted timber harvests, species habitats, watershed services, and climate stability (Myers, 1993a). Largely as a result of deforestation, tropical timber exports declined from US $8 thousand millions in 1986 to $6 thousand millions in 1991, and they are expected to fall to US $2 thousand millions in AD 2000 or shortly thereafter (Myers, 1991b). Twelve years ago, deforestation-derived flooding in India's Ganges Valley was levying costs of $1 thousand millions per annum (High Level Committee on Floods, Government of India, 1983).
- lost 60,000 square kilometres to desertification so severe that these lands will not be able to grow food for decades at best. Present desertification results in lost agricultural production worth US $42 thousand millions per year, and threatens the livelihoods of 850 million people (Kassas *et al.*, 1991; Dregne *et al.*, 1992; Chou & Dregne, 1993).
- caused the extinction of thousands of species (Raven, 1990; Wilson, 1992; Myers, 1994), some of which might have supplied new anti-cancer drugs such as the two from the Rosy Periwinkle (*Catharanthus roseus*). The cumulative market value of plant-based drugs in developed countries until the year 2000 is estimated to be $500 thousand millions (in 1984 dollars) (Principe, 1989, 1996).
- further depleted the stratospheric ozone shield in both the northern and southern hemispheres (Gleason *et al.*, 1993; *cf.* Polunin, 1993), causing it to lose still more of its capacity to protect us from cancer- and cataract-causing UV-B radiation, which also attacks crop-plants and marine food-chains.
- taken a further step towards a 'greenhouse-affected' world, which will cause profound environmental, economic, and political, disruptions throughout the global community (Intergovernmental Panel on Climate Change, 1992a, 1992b).

How far are environmental problems attributable to population growth? The most comprehensive and systematized analysis to date, with detailed documentation and methodical evaluation (Harrison, 1992), concludes that population growth in developing countries has accounted for:

- 72% of the expansion of arable lands during AD 1961–85, leading to desertification, deforestation, and deterioration of many natural environments;
- 69% of the increase in livestock numbers during AD 1961–85, leading to soil erosion, desertification, and methane (many times more active as a 'greenhouse' gas than carbon dioxide [CO_2]) emissions;

- 79% of the tropical deforestation during 1973–88, leading to thousands of species extinctions each year; and
- 46% of the growth in carbon dioxide emissions from fossil fuels during 1960–88, carbon dioxide being the gas that accounts for an estimated half of global-warming processes.

Several other analyses (*e.g.* Ehrlich & Ehrlich, 1990; Davis & Bernstam, 1991; Bongaarts, 1992; Falkenmark & Suprapto, 1992; Food and Agriculture Organization, 1993*a*, 1993*b*; Grainger, 1993; Ness *et al.*, 1993; Myers, 1993*c*, 1994; *cf.* Polunin & Burnett, 1993) have come up with similar findings, albeit with marginally different statistical conclusions. To reiterate a crucial caveat, many other factors are at work, such as negligent technology, poor land-use planning, economic inefficiencies, and faulty overall development policies. But so far as is feasible, these variables have been taken into account in the analyses cited. It is realistic, then, to offer the generalized conclusion that *population growth plays a prominent, and probably predominant, part in engendering environmental problems.*

The relationship operates the other way around as well. Environmental problems exacerbate population problems, as a declining resource-base depletes the livelihoods of societies whose economies are dependent on soils, water, forests, atmosphere, and climate. Of course this applies to all societies and economies; but especially does it apply to developing countries, with their greater reliance upon natural-resource stocks and their fewer alternatives for economic advancement, as well as with their fast-growing populations.

Particularly hard-pressed in the above connections are the 1.2 thousand million people living in absolute poverty; of these, 800 millions are chronically malnourished and 400 millions are semi-starving. *All of these totals have been increasing for more than a decade, with scant prospect of improvement* (World Bank, 1992). In several respects, these people appear to cause as much environmental damage as the rest of the developing world combined: they are primarily the ones who burn forests, desertify grasslands, and cultivate sloping croplands with resultant soil erosion (Little & Horowitz, 1987; Leonard *et al.*, 1989; Kates & Haarman, 1991; Mink, 1991). They are also the ones with the highest fertility rates, evidently a consequence of their poverty — which in turn is aggravated by environmental decline and population increase.

In sum, it may well be that we have achieved economic advancement in the past at environmental cost to the future's potential for still more advancement — and even at the more serious cost of an actual decline in human welfare, especially in the light of the ever-increasing demands of growing numbers of people.

The Green Revolution and Its Shortcomings

It is becoming apparent that 'Green Revolution' agriculture, its exceptional achievements notwithstanding, has entrained a number of covert costs through overloading of croplands leading to soil erosion, depletion of natural nutrients, and waterlogging/salinization of irrigation systems. These costs, while unnoticed or disregarded for decades, are now levying a price in terms of cropland productivity — precisely at a time when population growth remains high.

Soil erosion leads to an annual loss in grain output world-wide that is estimated at nine million tonnes; it also leads to other problems such as degradation of irrigated lands, accounting for the annual loss of another three million tonnes. In addition, there is pollution damage to crops worth another two million tonnes (Brown *et al.*, 1990). So the total toll from all forms of environmental degradation comes to 14 million tonnes of grain output lost per year. This is to be compared with gains from increased investments in irrigation, fertilizer, and other inputs, amounting in all to 29 million tonnes per annum (Brown *et al.*, 1990; *see also* Pimentel, 1993).

Thus environmental factors are causing the loss of almost half of all gains from technology-based and other advances in agriculture. Even worse, the proportion has risen steadily in recent years, and seems set to expand still further within the foreseeable future. The world simply cannot afford this continuing shortfall, as it requires an additional 28 million tonnes of grain each year to cater for the needs of population growth, let alone the demands of enhanced diets and other consequences of economic advancement.

Population Pressures in Developed Nations

Much population-associated environmental degradation — notably depletion of the agricultural resource-base — takes place in developing countries. But this is not to suggest that there are not important types of environmental degradation in developed countries as well, even though their population growth-rates are generally low. Britain, for example, with a 0.2% annual population growth-rate, features a net increase through natural growth (*i.e.* without immigration) of 116,000 persons per year. By contrast, Bangladesh, with a 2.4% annual population growth-rate, features a net increase of 2.7 million persons, which is 23 times as large.

However, because the energy consumption of each additional person in Britain is more than thirty times that of a Bangladeshi, Britain's population growth effectively contributes 3.9 times as much carbon dioxide to the global atmosphere, and hence (it is claimed) to global warming, as does Bangladesh's — yet Bangladesh stands to suffer far more through global warming than

Britain. The average British family comprises two children; but when we 'factor-in' resource consumption and pollution impacts, and then compare the British life-style with the global average, the 'real world' size of a British family is more like 10 children! Ironically, Britain could establish zero population growth by the simple expedient of eliminating unwanted births (Myers, 1994); also note that family planning benefits exceed costs by 5:1 (Laing, 1982; Estaugh & Wheatly, 1991).

A Super-pressure Emergent

All the above reflects upon the key concept of carrying capacity — a critical and controversial issue. Certain scientists, notably ecologists, assert that it is not only a key constraint to population growth, but that it can readily become an absolute factor. Others, notably economists, assert that carrying capacity is such a flexible affair — being, *inter alia*, subject to endless expansion through technology and policy interventions — that it soon ceases to have much operational value. Be this as it may, global carrying capacity can be defined as 'the number of people that the planet can support without irreversibly reducing its capacity to support people in the future' (Ehrlich *et al.*, 1989; *see also* Mahar, 1985; Daily *et al.*, 1993; Hardin, 1993).

While the above is a global-level definition, it applies also at the national level, albeit with many qualifications concerning international constraints such as trade, investment, and debt. Furthermore, it is a function of factors that reflect technological change, and supplies of food, water, and energy: also ecosystem services such as nutrient recycling, human capital, life-styles, social institutions, political structures, and cultural constraints — among many other factors, all of which interact with each other (Ehrlich & Ehrlich, 1990). Within the context of these variables and instances, there is emergent evidence that, in many instances, human numbers, with their consumption of resources *plus* the technologies deployed to supply that consumption and the pollutant impacts thereof, are already exceeding Earth's carrying capacity if humans are to continue to live at all equably.

For a specific illustration, consider the case of Kenya. Its 1993 population of 27 million people is projected to expand to 125 millions by the time, and if, zero growth of population is attained in the 22nd century. Yet even if the nation were to employ Western Europe's high-technology agriculture, it could support no more than 52 million people off its own lands (Food and Agriculture Organization, 1984; Harrison & Rowley, 1984) — and even if it were to achieve the two-child family forthwith, the population would still double because of demographic momentum (48% of present-day

Kenyans are under the age of 15, meaning that large numbers of future parents are already in place). So Kenya will have to depend on steadily increasing amounts of food from outside to support itself. But in large part because of its high population growth-rate — around 3% in recent years — its *per caput* economic growth has been well under 2% (World Bank, 1993).

Worse still is the circumstance that Kenya's terms of trade have been declining throughout the 1980s and early 1990s, until they are barely positive today — meaning that the country faces the prospect of diminishing financial reserves to purchase food abroad. Its export economy will need to flourish permanently, in a manner far better than it has ever achieved to date, if the country is to buy enough food to meet its fast-growing needs. Worst of all, the country will have to undertake this challenge with a natural-resource-base from which forests have almost disappeared, watershed flows for irrigation agriculture are often depleted*, and much topsoil has been eroded away (World Resources Institute, 1992).

Kenya thus shows many signs of already being in an 'overshoot' situation as concerns its carrying capacity. The best time to tackle the situation would have been during the far-back period when its population was barely starting to grow rapidly — and when all seemed well in terms of its capacity to feed itself for a while, though the source of its population/food dilemma was becoming entrenched. **Other nations with currently satisfactory capacity to ensure their food supplies should ponder Kenya's experience**. The main opportunity remaining for Kenya to relieve its situation lies with an immediate and vigorous effort to slow its population growth. Were the two-child family to be achieved in AD 2010 instead of the projected AD 2035, Kenya's ultimate population could be held to 72 millions, or 53 millions (42%) less than expected (McNamara, 1991*a*).

Environment: The Future Outlook and Potential Discontinuities

We should recognize, moreover, that the future may well feature a number of non-linear departures of environmental kinds, with all manner of repercussions for both our population and development prospects. An 'overshoot' outcome, as in Kenya's case (there are many other instances, *cf.* Ehrlich & Ehrlich, 1990; Harrison, 1992; and *see* below), can precipitate a decline in the capacity of environments to sustain human communities at their erstwhile level (Keyfitz, 1991*a*, 1991*b*; Myers, 1992*c*), whereupon we encounter a phenome-

---

* Here we recall vividly the account, given at our 4th ICEF (*cf.* Polunin & Burnett, 1993) by a distinguished Kenyan participant, of how, when he was a boy, he used to drink cool water from a spring near his abode. But now 'it is all dried-up'. — Ed.

non of environmental discontinuity. The latter phenomenon can arise when resource stocks or ecosystems have absorbed stresses over long periods without much outward sign of damage, until they suddenly reach a disruption level at which the cumulative consequences of stress reveal themselves in critical proportions. A familiar example is acid rain, better termed 'acidic precipitation'. We should anticipate that, as human communities continue to expand in numbers and consumptive activities, they will exert increasing pressures on already overburdened resource stocks and ecosystems, meaning that environmental discontinuities will become more and more common.

An instance has already arisen in the Philippines, where the frontier of agricultural expansion closed in the lowlands during the 1970s. As a result, multitudes of landless people started to migrate into the uplands, leading to a buildup of human numbers at a rate far greater than that of national population growth. The uplands contain the country's main remaining stocks of forests, and they feature much sloping land. The result was a marked increase in deforestation and a rapid spread of soil erosion. In other words, there occurred a 'breakpoint' in patterns of human settlement and environmental degradation. As long as the lowlands were less-than-fully occupied, it made little difference to the uplands whether there was 50% or 10% of space left. It was only when hardly any space at all was left that the situation altered radically; what had seemed acceptable became critical — and the profound shift occurred in a very short space of time (Myers, 1988).

The problem of land shortages is becoming widespread in many if not most developing countries, where land provides the livelihood for around three-fifths of the total population, and where the great bulk of the most fertile and most accessible land has already been taken (Harrison, 1992; Myers, 1994).

We encounter such non-linear relationships between resource exploitation and population growth with respect to many other resource stocks, notably water supplies, forests, fisheries, soil cover, and pollution-absorbing services of the atmosphere (Repetto, 1987; Dasgupta, 1992; Keyfitz, 1992; Myers, 1992c, 1992d). Whereas resource exploitation may have been expanding gradually for long periods without undue harm, the switch in the scale of exploitation, induced through a phase of rapid population growth, can readily result in a slight initial exceeding of the sustainable yield, whereupon the process of resource depletion is precipitated with surprising rapidity and often with extreme human deprivation.

Global Warming and Agriculture

The biggest discontinuity in the foreseeable future may well be climate change in the form of global warming. Except by some cautioners (*e.g.* Bryson, 1993), the process is expected to cause a pronounced shift in climate systems world-wide; and so far as we can discern, it will exert profoundly harmful impacts on the world's, and especially developing countries', capacity to grow food (Schneider, 1989; Intergovernmental Panel on Climate Change, 1992*a*, 1992*b*). It is calculated to cause a 9–11% decline in grain harvests, *plus* substantial shortfalls for many other crops, in developing countries, together with reduced yields in certain 'bread-baskets' of the temperate zones; and the consequences could include a rise in the number of severely undernourished people to an eventual total of one thousand millions (Rosenzweig *et al.*, 1993). Before any serious global warming sets in, moreover, increased droughts (whether natural or human-made) could cause a 10% reduction in grain harvests on the average three times per decade, resulting in the malnutrition deaths of as many as 800 million people over a period of two decades (Daily & Ehrlich, 1990).

As a further result, there is likely to be a steep increase in the number of 'environmental refugees', reaching as many as 150 millions within a few decades, or six times the total of all refugees nowadays (Myers, 1993*d*). Each of the analyses cited concludes that *the most productive and readily available mode of adaptation to the global-warming threat would be to reduce population growth forthwith*. In the global-warming scenario (Rosenzweig & Parry, 1992), a low rate of population growth would reduce the number of people at risk by almost half.

When we consider all environmental disruptions together, in conjunction with the compounding impacts of population growth, there are potentially a multitude of discontinuities, often with synergistic interactions, that will generate markedly adverse impacts. There is good reason, then, for us to prepare ourselves for a greater environmental crisis, arriving more rapidly, than has commonly been anticipated.*

Sustainable Development†: The Ultimate Imperative

As has been emphasized on numerous occasions (*e.g.* World Commission on Environment and Development, 1987; United Nations Conference on Environment and Development, 1992), there

---

* Especially in view of the contingency that climate change might be far more rapid and extreme than is generally thought likely — *see e.g.* the article entitled 'Finding on climate alarms scientists' by Walter L. Sullivan (*International Herald Tribune* [Zürich] Nr 34,331, pp. 1 and 3, 16 July 1993). — Ed.

† This seeming ideal becomes less of a mere catch-phrase if preceded (at least mentally) always by 'environmentally' or 'ecologically' as is presumably understood by this distinguished Author to preclude economists' and industrialists' understanding of 'development'. We have done this in editing *Environmental Conservation* for years and have noticed latterly that some leading authorities and major concerns are doing likewise. — Ed.

can be little advance on any front except within an overall context of sustainable development[†]. This major departure can be defined as 'development that meets the needs of the present without compromising the ability of future generations to meet their own needs' (World Commission on Environment and Development, 1987).

Regrettably, no country is anywhere near approaching sustainable development. By way of brief illustration, note that subsidies in many economic sectors serve to undermine the environmental resource-base and thus undermine long-term development as well. Subsidies for over-productive agriculture in developed nations total US $350 thousand millions per annum; and for fossil fuels worldwide, they total $300 thousand millions, which is often ten times as great as subsidies for non-polluting-energy alternatives. Still other subsidies foster misuse or over-use of water and forest resources, and also of pesticides. These figures throw light on the budget proposed at the Rio 'Earth Summit' for the *Agenda 21 Action Plan* (United Nations Conferences on Environment and Development, 1992), namely $625 thousand millions per annum — a figure that has been viewed as quite unattainable.

For more details on the need and opportunity to promote sustainable development, and use of the term, *see* the last side-headed section of this chapter, before the brief Conclusion.

POPULATION, ENVIRONMENT, AND DEVELOPMENT: OUR CHOICES

Strategies for Reducing Population Growth

There is much scope for us to expand our efforts to reduce population growth. As already noted, socio-economic advancement in developing countries often fosters motivation for family planning in addition to many other benefits. The same applies to improvement in the status of developing-country women, particularly as concerns education. Girls receive little more than half as much schooling as boys, and the female literacy rate is less than 70% of that of males (United Nations Development Programme, 1992). But women with four years of secondary schooling *reduce their fertility by an average of two children.*

To educate all girls as much as boys in low-income countries (these being the countries where women are most disadvantaged and where fertility rates are highest) would cost $2.4 thousand millions a year, or less than one-quarter of one per cent of these countries' collective GDP (Summers, 1992; *see also* Kaul & Menon, 1993). Kenya now has one of the highest female literacy rates in the whole of Africa, and this, together with a recent rapid fall in infant mortality, is thought to lie behind the decline in the country's population growth-rate from 4.1% in 1986 to 2.7% in 1996 — a breakthrough

---

[†] *See* lower footnote on preceding page. — Ed.

that may soon be matched by countries as economically and socio-culturally disparate as Zimbabwe, Botswana, Cameroon, and Senegal (Sinding, 1993).

We face a further exceptional opportunity with respect to child mortality. No country has attained a low fertility-rate while its child mortality-rate has remained high (Grant, 1993; United Nations Population Fund, 1993). During the 1990s we could save 100 million children, or three-quarters of those at risk, through existing simple technologies to counter diarrhoea and disease, at a cost of little over $400 millions a year. Even if rich countries were to cover the entire cost (they currently pay around one-quarter of it), it would work out at only US $1.33 per annum for each of the estimated 300 million taxpayers in the developed world.

Most important of all is the need and opportunity to expand family-planning services. Family planning has already been respon-sible for fully one-half of the decline in fertility, from 6.4 to 3.4 children, in developing countries since the 1950s; and today family planning eliminates the births of 43 million unwanted children (ac-tual births being some 140 millions world-wide) per annum (Bon-gaarts, 1993; *see also* Freedman, 1990; Phillips & Ross, 1992). The demand for family planning is now greater than ever. Half of developing-country births are unplanned and one-quarter are un-wanted. Fortunately, more than 90% of developing-world people live in countries where population growth is now officially consi-dered too high for economic advancement; accordingly it should be reduced with all due dispatch. To meet these aspirations, we shall need (among other measures) to double our outlays on family planning by the year 2000. Fortunately these outlays are minor, and not only when compared with the costs of an overpopulated world: the developed countries' share will work out at one penny (or US cent) per day per citizen (Potts, 1990) — a sound investment to help ensure that every child is a wanted child.*

Particularly pertinent are the needs of the estimated 300 million women who possess motivation to limit their fertility but lack birth-control facilities.† Were these needs to be met (as they should be on humanitarian grounds alone), it would result, according to a very conservative estimate (Sinding, 1992), in a decline from roughly four to just over three children per reproductive woman. In turn, this would represent a decline of 50% on the way from present-level fertility to replacement-level fertility. It would also reduce the even-

---

* Another advantage could be reduction in the incidence of future dictatorship, a high proportion of the world's most notorious and cruel dictators having apparently been unwanted children. — Ed.

† For illustrated accounts of this aspect and its hopeful significance, *see* Chapter 14. — Ed.

tual global population by at least two thousand million people (Bongaarts, 1990). At a cost of $15 per couple per year, the budget need be no more than US $4.5 thousand millions.

Furthermore, it would bring major benefits to women in that it would cut back on pregnancy-related problems — including the fast-increasing number of abortions. Each year at least 500,000 developing-country women die of pregnancy-related problems (several times more have narrow escapes), or a total greater than the AIDS toll of women and men combined (World Health Organization, 1992; Zahr & Royston, 1993).

To cite James P. Grant, Executive Director of the United Nations Children's Fund (UNICEF), 'Family planning could bring more benefits to more people at less cost than any other single technology now available to the human race. But it is not appreciated widely enough that this would still be true even if there were no such thing as a population problem' (Grant, 1993). In fact, family planning costs would be far outweighed by benefits. In Mexico, every peso spent on family planning for the urban population during the period 1972–84 generated a saving, in terms of infrastructure services not required (schools, health facilities, housing, employment, etc.), of four to five pesos (Nortman et al., 1986). A similar analysis in Thailand for the period 1972–80 showed a savings ratio of 7 to 1 (Chao & Allen, 1984). Note, too, that in the United States every $1 of public funds spent on contraceptive services generates health benefits worth $4.40 (Forrest & Singh, 1990).

The chief constraint is time — and the present decade is critical for the long-term future. If the two-child family becomes the norm in developing countries early next century, global population should eventually cease to grow at a level below 9 thousand millions. If, instead, fertility declines no lower than 2.5 children, global population is expected to grow to 19 thousand millions by the year 2100, and to keep on expanding steadily thereafter. In a longer-term perspective, suppose that developing-country fertility eventually settles at round-about 2.06, that 'round-about' is critical. If it were only 5% higher or lower, the consequences would be almost a fourfold difference by the year 2150, namely 5.6 thousand millions versus 20.7 thousand millions (United Nations, 1992).

To illustrate the key factor of demographic inertia with respect to a specific country, consider the prospect for Pakistan. In 1990 its population was 115 millions with a growth-rate of 3.2%. If the country had then taken steps to achieve replacement fertility by AD 2010 instead of the anticipated AD 2040, it would have held its eventual population total to 334 millions, or 222 millions fewer than the projected 556 millions (McNamara, 1991a). (As a measure of what awaits Pakistan with its population pressures, note that 90% of the country's food production depends upon irrigation agriculture; farmers already divert 76% of the flows of the River Indus, being by far Pakistan's largest river-system, and 89% of recharge of ground-water stocks [Qutub, 1993].)

Fortunately, it is far from impossible for a country to achieve replacement fertility within 20 years. Thailand reduced its fertility from 6.5 in 1969 to 2.1 in 1989, and a number of other countries — such as China, South Korea, and Taiwan, also Java in Indonesia and Kerala State in India — have achieved almost as rapid a rate of fertility decline. Moreover, Pakistan features a level of unmet family-planning needs at 20%.

Strategies for Reducing Environmental Degradation

There is much scope, too, for reducing environmental degradation. Most of the specific measures have been set out in the UNCED Action Plan known as *Agenda 21* (United Nations Conference on Environment and Development, 1992), accepted in principle at the Rio 'Earth Summit', with a budget for the most pressing international problems totalling $625 thousand millions per annum. In practice, however, little has been done to implement these measures on the grounds that they are too expensive in the current economic climate — though little mention is made of the concealed costs of inaction, which are often several times as large as the costs of action.

As examples we may note that water shortages in developing countries cause diseases that, through work-days lost to sickness, levy a cost of $125 thousand millions a year (Pearce, 1993); with this we should contrast the cost of supplying both water and sanitation facilities, namely $50 thousand millions per annum (Christmas & Rooy, 1991). Again, desertification leads to lost agricultural output worth $42 thousand millions per year, whereas an anti-desertification programme would cost between $10 thousand millions and $22 thousand millions per annum (United Nations Environment Programme, 1991; Chou & Dregne, 1993).

Similar positive payoffs apply to efforts to counter the fuel-wood, deforestation, biodiversity, and global warming, crises (Ehrlich & Ehrlich, 1996; Myers, 1996). The most notable example lies with the stratospheric ozone-shield depletion: for the United States to phase out ozone-destroying chemicals (the nation accounts for one-third of the world's production and consumption) will cost $36 thousand millions, largely by the year 2000, compared with costs of lives otherwise lost to skin cancers during the next one-and-three-quarters centuries, totalling some $40 billions (US trillions) (IFC Inc., 1992).

Many opportunities are available to safeguard environmental resources in individual sectors. Tropical forestry, for instance, is subject to a host of subsidies that stimulate overexploitation (Repetto & Gillis, 1988). These subsidies have resulted in an outburst of cattle ranching in Brazilian Amazonia, where commercial enterprises have made inordinate profits at a cost to the national economy, through loss of commercial timber alone, of $2.5 thousand millions annually (Repetto, 1990). Nor is forest overexploitation a phenomenon confined to tropical countries. One of the rain-forests that is being most rapidly depleted is not in Amazonia or Borneo but the Tongass National Forest in Alaska, USA, due primarily to subsidies on the part of the US Government. Similarly 'perverse' subsidies foster massive misuse and over-use of water, fertilizer, pesticides, etc. (Repetto, 1986; Faeth *et al.*, 1991; Pimentel, 1993), and energy (Kosmo, 1987; Holdren, 1991; Romm & Lovins, 1992) — all of them with adverse repercussions of both environmental and economic kinds. In total, these perverse subsidies may now amount to $600 thousand millions a year world-wide — a remarkable proportion of the global economy worth around $26 US trillions annually.

There are many other modes of environmental protection at the macro-level of national planning. A notable instance is natural resource accounting in order to reflect the environmental underpinnings of economic activity, and thus serving as a corrective to the 'blunt-instrument' style of accounting in the form of Gross National Product (Daly & Cobb, 1989; Repetto *et al.*, 1989; Peskin & Lutz, 1990). An allied planning strategy lies with correction of pricing policies, especially to reflect externalities (Pearce *et al.*, 1989, 1991). Were the price of gasoline ('petrol') in the United States to internalize the costs of urban smog, acid rain, low-level ozone pollution, and global warming, *plus* the expense of securing Persian Gulf oil (as well as to reflect the many subsidies in support of the petroleum and automobile industries), the 'true' price would be at least four times the present price (Romm, 1992; Romm & Lovins, 1992).

Strategies for Sustaining Development

All the measures listed under the last heading would contribute to the imperative of environmentally* sustainable development for all nations, whether developed or developing. Many other measures are presented in the *Agenda 21* Action Plan adopted at the Rio 'Earth Summit' — a global gathering which was designated a conference on both environment and development, being entitled the United Nations Conference on Environment and Development (UNCED). In any case these measures are sufficiently well known for there to be no need to repeat them here.

---

* *See* lower footnote on p. 32. — Ed.

But consider some of the particular challenges of developing countries, and how much can be accomplished through better-targeted funding for development than is currently practised. Aid-donor countries could double the proportion of Official Development Assistance (ODA) which they direct to the poorest countries, and which is now only one-quarter of what they give. They could likewise direct special emphasis to the poorest 20% of the global population — the 1.2 thousand million people who receive only 1.4% of global income, 1.3% of global investment, and 0.2% of global commercial-bank lending — and who feature the highest and most intractable fertility rates (Haq, 1992; United Nations Development Programme, 1993).

Still more to the point, developing countries and aid-donor countries alike devote only around 10% of their expenditures to priority human needs such as nutrition, primary health-care, water and sanitation, basic education, and family planning. Yet the communities which are most lacking in these essentials are the 1.2 thousand millions living in absolute poverty. The challenge could be largely surmounted if the funding proportion were to be doubled to 20%. At the same time, this would make a marked contribution towards enhancing the motivation for reduced fertility. All in all, the task could be accomplished for a mere $25 thousand millions a year, of which only one-third need come from developed countries (Grant, 1993). By way of illustration, the additional cost for the United States would be $2 thousand millions, or a mere $7.5 per American. Alternatively, the sum could easily be made available by simply restructuring the US aid programme (including military assistance), now totalling some $9 thousand millions per annum.

In addition, there is much more that developing countries can do to help themselves. If they lowered their military expenditures to AD 1970 levels, privatized public enterprises, corrected distortive development policies (such as perverse subsidies), rooted out corruption, and generally improved national governance, they would readily release $50 thousand millions per annum (United Nations Development Programme, 1991). They could also mobilize substantial funds through a reordering of priorities within individual sectors. In the public-health field, for instance, they could shift the budgetary emphasis from curative to preventive medicine — a measure that alone would take care of basic health needs for all citizens — and thus indirectly promote the motivation for family planning.

Further sustainable-development strategies would not be expensive for developing countries, while also assisting the population cause. To eliminate deaths from famine would cost $0.55 thousand millions per year; to cut malnutrition among women and children, $1.6 thousand millions; and to reduce hunger among the poorest households, $6.4 thousand millions. The total cost of these measures would be $8.55 thousand millions, or just $7.13 for each of the 1.2 thousand million people living in absolute poverty; and remember that these are the people who have the largest families (Chen & Emlen, 1992; *see also* Kasperson & Kates, 1990). Note meanwhile that developing countries currently spend an average of around US $40 per citizen per annum on military activities.

Much could be done, too, to reduce the debt-burden on developing countries. They paid $1.4 thousand billions (American trillions) between 1982 and 1991 merely to service their debt, which meantime expanded by almost one-half (World Bank, 1992). During the late 1980s, and largely due to external debt, there was a net financial flow, from the South to the North, of the order of $40 thousand millions a year, *i.e.* all official aid, loans, and investments, from the North, were more than cancelled out by debt payments. This situation has now been relieved to some extent: but it still leaves developing countries with scant net income in terms of their financial relationships with developed countries.

Consider still another aspect of North–South relationships within the development context. Were trade liberalization to be extended by developed countries to developing countries (to date, free-trade operates mainly among developed countries), this would enable developing countries to earn an extra $100 thousand millions per annum through export revenues, or almost twice as much as all the aid they receive from developed countries. Because of their unequal or restricted access to global financial and labour markets as well as to world trade, developing countries are deprived of some $500 thousand millions per year (Haq, 1992).

The finest opportunity for creative action probably lies with the putative 'Peace Dividend' in the wake of the Cold War's end. A range of estimates (McNamara, 1991b; Sivard, 1991; Strong, 1992) proposes that reduced defence etc. spending — on the part of developing as well as developed countries — could release an average of $120 thousand millions a year during the 1990s. This would enable the global community to do far more than currently to address global problems of population, environment, and development. The funds, formerly deployed on conventional security, would thereby help the global community to purchase a greater degree of all-around and enduring security (Myers, 1993c).

CONCLUSION

The triad problems of population growth, environmental decline, and unsustainable development, are generating so much momentum that they threaten to overwhelm Humankind's and Nature's prospects for an acceptable future. Fortunately there is abundant scope, though probably not very much time, for us to formulate response strategies to confront the challenges which they represent. There is thus an opportunity to turn unprecedented problems into unprecedented opportunities — to develop a sustainable Homosphere within a durable Biosphere (Polunin, in press).

Correct action should occupy the best scientists, activators, and political leaders, well into the future, and (to quote the last-named Author (*ibid.*) 'offer our species a fine (and perhaps final!) chance of justifying at last the specific epithet of *sapiens*'.

REFERENCES

ABERNETHY, V.D. (1993). *Population Politics: The Choices That Shape Our Future.* Insight Books/Plenum Press, New York, NY, USA: xix + 350 pp.

BONGAARTS, J. (1990). The measurement of wanted fertility. *Population and Development Review*, **16**, pp. 487–506.

BONGAARTS, J. (1992). Population growth and global warming. *Population and Development Review*, **18**(2), pp. 299–319.

BONGAARTS, J. (1993). *The Fertility Impact of Family Planning Programs.* The Population Council, New York, NY, USA: 26 pp.

BRAUN, J. VON & PAULINO, L. (1990). Food in Sub-Saharan Africa: trends and policy challenges for 1990s. *Food Policy*, **15**, pp. 505–17.

BROWN, L.R., DURNING, A., FLAVIN, C., FRENCH, H., JACOBSON, J., LOWE, M., POSTEL, S., RENNER, M., STARKE, L. & YOUNG, J. (1990). *State of the World 1990.* W.W. Norton, New York, NY, USA: xvi + 253 pp., illustr.

BROWN, L.R., FLAVIN, C. & KANE, H. (1992). *Vital Signs 1992: The Trends That Are Shaping Our Future.* Worldwatch Institute, Washington, DC, USA: 131 pp., illustr.

BROWN, L.R., DURNING, A., FLAVIN, C., FRENCH, H., JACOBSON, J., LENSSEN, N., LOWE, M., POSTEL, S., RENNER, M., STARKE, L., WEBER, P. & YOUNG, J. (1993). *State of the World 1993.* W.W. Norton, New York, NY, USA: xix+ 268 pp., illustr.

BROWN, L.R., FLAVIN, C., & FRENCH, H. (and others) (Ed. L. STARKE) (1997). *State of the World 1997.* W.W. Norton & Co., New York & London: xvii + 229 pp., figs & tables.

BRYSON, R.A. (1993). Simulating past and forecasting future climates. *Environmental Conservation*, **20**(4), pp. 339–46, 6 figs.

CHAO, D. & ALLEN, K.B. (1984). A cost–benefit analysis of Thailand's family planning programme. *International Family Planning Perspectives*, **10**, pp. 75–81.

CHEN, R.S. & EMLEN, J.S. (Eds) (1992). *Halving World Hunger by the Year 2000.* World Hunger Program, Brown University, Providence, Rhode Island, USA: [not available for checking].

CHOU, N.-T. & DREGNE, H.E. (1993). Desertification control: cost/benefit analysis. *Desertification Control Bulletin*, **22**, pp. 20–36.

CHRISTMAS, J. & ROOY, C. DE (1991). The Water Decade and beyond. *Water International*, **16**, pp. 127–34.

DAILY, G.C. & EHRLICH, P.R. (1990). An exploratory model on the impact of rapid climate change on the world food situation. *Proceedings of Royal Society of London B*, **241**, pp. 232–44.

DAILY, G.C., EHRLICH, A.H. & EHRLICH, P.R. (1993). *Optimum Human Population Size.* Paper for Optimum Population Conference, Cambridge, UK, 9–11 August 1993: [not available for checking].

DALY, H.E. (1991). Elements of environmental macroeconomics. Pp. 32–46 in *Ecological Economics* (Ed. R. COSTANZA). Columbia University Press, New York, NY, USA: xiii + 525 pp., illustr.

DALY, H.E. & COBB, J.B., JR (1989). *For the Common Good: Redirecting the Economy Toward Community, the Environment and a Sustainable Future.* Beacon Press, Boston, Massachusetts, USA: viii + 482 pp., illustr.

DASGUPTA, P. (1992). Population, resources, and poverty. *Ambio,* 21(1), pp. 95–101.

DAVIS, K. & BERNSTAM, M.S. (Eds) (1991). *Resources, Environment, and Population: Present Knowledge, Future Options.* Oxford University Press, New York, NY, USA: xii + 421 pp., illustr.

DREGNE, H., KASSAS, M. & ROSANOV, B. (1992). A new assessment of the world's status of desertification. *Desertification Control Bulletin,* 20, pp. 6–18.

EHRLICH, P.R. & EHRLICH, A.H. (1990). *The Population Explosion.* Simon & Schuster, New York, NY, USA: 320 pp.

EHRLICH, P.R. & EHRLICH, A.H. (1991). *Healing the Planet: Strategies for Resolving the Environmental Crisis.* Addison-Wesley Publishing Co. Inc., Menlo Park, California, USA: xv + 366 pp.

EHRLICH, P.R. & EHRLICH, A.H. (1996). *Betrayal of Science and Reason: How Antienvironmental Rhetoric Threatens our Future.* Island Press, Washington, DC, USA: [not available for checking].

EHRLICH, P.R. & HOLDREN, J.P. (Eds) (1988). *The Cassandra Conference: Resources and the Human Predicament.* Texas A&M University Press, College Station, Texas, USA: xi + 330 pp.

EHRLICH, P.R., DAILY, G.C., EHRLICH, A.H., MATSON, P. & VITOUSEK, P. (1989). In *Global Change and Our Common Future* (Eds R. DEFRIES & T. MALONE). National Academy Press, Washington, DC, USA: [not available for checking].

ESTAUGH, V. & WHEATLY, J. (1991). *Family Planning and Family Well-being.* Family Policy Studies Centre, London, England, UK: 11 pp.

FAETH, P., REPETTO, R., KROLL, K., DAI, Q & HELMERS, G. (1991). *Paying the Farm Bill: US Agricultural Policy and the Transition to Sustainable Agriculture.* World Resources Institute, Washington, DC, USA: ix + 70 pp.

FALKENMARK, M. & SUPRAPTO, R.A. (1992). Population–landscape interactions in development: a water perspective to environmental sustainability. *Ambio,* 21(1), pp. 31–6.

FOOD & AGRICULTURE ORGANIZATION (1984). *Potential Population Supporting Capacities of Lands in the Developing World.* Food and Agriculture Organization, Rome, Italy: [not available for checking].

FOOD & AGRICULTURE ORGANIZATION (1993a). *Forest Resources Assessment for the Tropical World.* Food and Agriculture Organization, Rome, Italy: 14 pp.

FOOD & AGRICULTURE ORGANIZATION (1993b). *Agriculture 2010.* Food & Agriculture Organization, Rome, Italy: [not available for checking].

FORREST, J.D. & SINGH, D. (1990). Public sector savings resulting from expenditures for contraceptive services. *Family Planning Perspectives,* 22(1), pp. 6–15.

FREEDMAN, R. (1990). Family planning programs in the Third World. *Annals of the American Academy of Political and Social Science.* Special Issue, 510. [Not available for checking.]

GLEASON, J.F. & 13 others (1993). Record low global ozone in 1992. *Science,* 260, pp. 523–5.

GRAINGER, G. (1993). *Controlling Tropical Deforestation.* Earthscan Publications, London, England, UK: [not available for checking].

GRANT, J.P. (1992). *State of the World's Children, 1992.* Oxford University Press, New York, NY, USA: 128 pp., illustr.

GRANT, J.P. (1993). *Children and Women — The Trojan Horse Against Mass Poverty?* UNICEF, New York, NY, USA: 11 pp.

HAQ, M. (1992). *Human Development in a Changing World.* Human Development Report Office, United Nations Development Programme, New York, NY, USA: 4 pp.

HARDIN, G. (1993). *Living Within Limits: Ecology, Economics, and Population Taboos.* Oxford University Press, New York, NY, USA: [not available for checking].

HARRISON, P. (1992). *The Third Revolution: Environment, Population and a Sustainable World.* I.B. Tauris & Co. Ltd., London, England, UK: xi + 359 pp., illustr.

HARRISON, P. & ROWLEY, J. (1984). *Human Numbers, Human Needs.* International Planned Parenthood Federation, London, England, UK: 64 pp., illustr.

HIGH LEVEL COMMITTEE ON FLOODS, GOVERNMENT OF INDIA (1983). *Report on the Emergent Crisis.* High Level Committee on Floods, Government of India, New Delhi, India: [not available for checking].

HOLDREN, J.P. (1990). Energy in transition. *Scientific American,* **263**(3), pp. 108–15.

HOLDREN, J.P. (1991). Population and the energy problem. *Population and Environment,* **12**, pp. 231–55.

IFC INC. (1992). *Regulatory Impact Analysis: Compliance with Section 604 of the Clean Air Act for the Phaseout of Ozone Depleting Chemicals: Addendum.* (Prepared for the Global Change Division, US Environmental Protection Agency.) IFC Inc., Washington, DC, USA: [not available for checking].

INTERGOVERNMENTAL PANEL ON CLIMATE CHANGE (1992a). *Climate Change 1992: The Supplementary Report to the IPCC Scientific Assessment* (Eds J.T. HOUGHTON, B.A. CALLENDER & S.K. BARBEY). Cambridge University Press, New York, NY, USA: xii + 200 pp., illustr.

INTERGOVERNMENTAL PANEL ON CLIMATE CHANGE (IPCC) (1992b). *Global Climate Change and the Rising Challenge of the Sea.* IPCC Coastal Zone Management Subgroup, Rijkswaterstraat, Netherlands: [not available for checking].

KASPERSON, J. & KATES, R.W. (Eds) (1990). Overcoming hunger in the 1990s. *Food Policy,* **15**(4), pp. 274–6.

KASSAS, M., AHMAD, Y. & ROSANOV, B. (1991). Desertification and drought: an ecological and economic analysis. *Desertification Control Bulletin,* **20**, pp. 19–29.

KATES, R.W. & HAARMAN, V. (1991). *Poor People and Threatened Environment: Global Overviews, Country Comparisons, and Local Studies.* World Hunger Program, Brown University, Providence, Rhode Island, USA: 70 pp.

KAUL, I. & MENON, S. (1993). *Human Development: From Concept to Action.* Human Development Report Office, United Nations Development Programme, New York, NY, USA: 25 pp.

KENDALL, H. & PIMENTEL, D. (1993). Constraints on the expansion of the global food supply. *BioScience,* [not available for checking].

KESSLER, E. (Ed.). (1992). *Population, Natural Resources and Development. Ambio* Special Issue, **21**(1), 123 pp.

KEYFITZ, N. (1991a). Population growth can prevent the development that would slow population growth. Pp. 39–77 in *Preserving the Global Environment: The Challenge of Shared Leadership* (Ed. J.T. MATHEWS). W.W. Norton, New York, NY, USA: 362 pp.

KEYFITZ, N. (1991b). *The Impact of Population Growth on the Physical Environment.* International Institute for Applied Systems Analysis, Laxenburg, Austria: [not available for checking].

KEYFITZ, N. (1992). Completing the worldwide demographic transition: the relevance of past experience. *Ambio,* **21**(1), pp. 26–30.

KOSMO, M. (1987). *Money to Burn? The High Costs of Energy Subsidies.* World Resources Institute, Washington, DC, USA: vii + 68 pp., illustr.

LAING, W.A. (1982). *Family Planning: The Benefits and Costs.* Policy Studies Institute, London, England, UK: 16 pp.

LEONARD, H.J., BROWDER, J.O., CAMPBELL, T., BOER, A.J. DE, JOLLY, A., STRYKER, J.D. & YUDELMAN, M. (1989). *Environment and the Poor: Development Strategies for a Common Agenda*. Transaction Publishers, New Brunswick, New Jersey, USA: x + 222 pp., illustr.

LITTLE, P.D. & HOROWITZ, M.M. (Eds) (1987). *Lands at Risk in the Third World: Local-level Perspectives*. Westview Press, Boulder, Colorado, USA: [not available for checking].

MAHAR, D. (Ed.) (1985). *Rapid Population Growth and Human Carrying Capacity*. World Bank Staff Working Paper No. 690, World Bank, Washington, DC, USA: [not available for checking].

MCNAMARA, R.S. (1990). Population and Africa's development crisis. *Populi*, **17**(4), pp. 35–43.

MCNAMARA, R.S. (1991a). *A Global Population Policy to Advance Human Development in the 21st Century*. Rafael M. Salas Memorial Lecture, United Nations Population Fund, New York, NY, USA: 56 pp., illustr.

MCNAMARA, R.S. (1991b). Toward a new world order. *EcoDecision*, **2**, pp. 14–9.

MINK, S. (1991). *Poverty, Population, and the Environment*. The World Bank, Washington, DC, USA: 46 pp.

MYERS, N. (1988). Environmental degradation and some economic consequences in the Philippines. *Environmental Conservation*, **15**(3), pp. 205–14.

MYERS, N. (1991a). The world's forests and human populations: the environmental interconnections. Pp. 237–51 in *Resources, Environment, and Population: Present Knowledge, Future Options* (Eds K. DAVIS & M.S. BERNSTAM). Oxford University Press, New York, NY, USA: xii + 421 pp., illustr.

MYERS, N. (1991b). *Population, Resources and the Environment: The Critical Challenges*. Banson Books, London, for United Nations Population Fund, New York, NY, USA: vi + 154 pp., illustr.

MYERS, N. (1992a). *The Environmental Consequences for the European Community of Population Factors Worldwide and Within the Community*. European Commission, Brussels, Belgium: 124 pp.

MYERS, N. (1992b). *The Primary Source: Tropical Forests and Our Future* (expanded edition). W.W. Norton, New York, NY, USA: xxxii + 416 pp., illustr.

MYERS, N. (1992c). Population/environment linkages: discontinuities ahead. *Ambio*, **21**(1), pp. 116–8.

MYERS, N. (Ed.) (1992d). *Tropical Forests and Climate*. Kluwer Academic Publishers, Dordrecht, Netherlands: vi + 265 pp., illustr.

MYERS, N. (1993a). Tropical forests: the main deforestation fronts. *Environmental Conservation*, **20**(1), pp. 9–16, map and tables.

MYERS, N. (1993b). The question of linkages in environment and development. *BioScience*, **43**(5), pp. 302–10.

MYERS, N. (1993c). *Ultimate Security: The Environmental Basis of Political Stability*. W.W. Norton, New York, NY, USA: xi + 308 pp., illustr.

MYERS, N. (1993d). Environmental refugees: how many ahead? *BioScience*, **43**(11), pp. 752–61.

MYERS, N. (1994). Chapters in *Scarcity or Abundance: A Debate on the Environment* (N. MYERS & J. SIMON). W.W. Norton, New York, NY, USA: [not available for checking].

MYERS, N. (1996). *Ultimate Security: The Environmental Basis of Political Stability*. Island Press, Washington, PC, USA: [not available for checking].

NESS, G.D., DRAKE, W.D. & BRECHIN, S.R. (Eds) (1993). *Population–Environment Dynamics: Ideas and Observations*. University of Michigan Press, Ann Arbor, Michigan, USA: xv + 456 pp., illustr.

NORTMAN, D.L., HALVAS, J. & RAMBAGO, A. (1986). A cost–benefit analysis of the Mexican social security administration's family planning program. *Studies in Family Planning*, **17**, pp. 1–6.

PEARCE, D.W. (1993). *Economic Values and the Natural World.* Earthscan Publications, London, England, UK: xi + 129 pp., illustr.

PEARCE, D.W. & ATKINSON, G. (1992). *Are National Economies Sustainable? Measuring Sustainable Development.* Centre for Social and Economic Research on the Global Environment, University College, London, England, UK: [not available for checking].

PEARCE, D.W., MARKANDYA, A. & BARBIER, E.B. (1989). *Blueprint for a Green Economy.* Earthscan Publications, London, England, UK: xvi + 192 pp., illustr.

PEARCE, D.W., BARBIER, E.B., MARKANDYA, A., BARRETT, S., TURNER, R.K. & SWANSON, T. (1991). *Blueprint 2: Greening the World Economy.* Earthscan Publications, London, England, UK: 232 pp., illustr.

PESKIN, H.M. & LUTZ, E. (1990). *A Survey of Resource and Environmental Accounting in Industrialized Countries.* The World Bank, Washington, DC, USA: 74 pp.

PHILLIPS, J.F. & ROSS, J.A. (Eds) (1992). *Family Planning Programmes and Fertility.* Clarendon Press, Oxford, England, UK: [not available for checking].

PIMENTEL, D. (Ed.) (1993). *World Soil Erosion and Conservation.* Cambridge University Press, Cambridge, England, UK: xii + 349 pp., illustr.

PINSTRUP-ANDERSEN, P. (1992). *Global Perspectives for Food Production and Consumption.* International Food Policy Research Institute, Washington, DC, USA: 24 pp.

POLUNIN, N. (1993). Obscure can be supreme: a double lesson from the stratospheric ozone shield. *Environmental Conservation,* **20**(2), p. 97.

POLUNIN, N. (in press). The imperatives of Biosphere conservation and wise Homosphere development. (Invited Essays in Honour of Clement A. Tisdell.) *International Journal of Social Economics.*

POLUNIN, N. & BURNETT, SIR J. (Eds) (1993). *Surviving With The Biosphere: Proceedings of the Fourth International Conference on Environmental Future (4th ICEF), held in Budapest, Hungary, during 22–27 April 1990.* Edinburgh University Press, Edinburgh, Scotland, UK: xxii + 572 pp., illustr.

POTTS, D. (1990). *A Penny a Day.* Family Health International, Research Triangle Park, North Carolina, USA: 16 pp.

PRINCIPE, P. (1989). The economic significance of plants and their constituents as drugs. Pp. 1–17 in *Economic and Medicinal Plant Research,* Vol. 3 (Eds H. WAGNER, H. HIKINO & N. FARNSWORTH). Academic Press, London, England, UK: [not available for checking].

PRINCIPE, P. (1996). Monetizing the pharmacological benefits of plants. Pp. 191–218 in *Tropical Forest Medical Resources and the Conservation of Biodiversity* (Eds M.J. BALICK *et al.*). Columbia University Press, New York, NY, USA: [not available for checking].

PURCELL, J.M. (1993). *Towards a Comprehensive Approach to the Migration Challenges of the 1990s.* Office of the Director-General, International Organization for Migration, Geneva, Switzerland: 8 pp.

QUTUB, S.A. (1993). Pakistan on collision course. *People and the Planet,* **2**(2), pp. 24–5.

RAVEN, P.R. (1990). The politics of preserving biodiversity. *BioScience,* **40**, pp. 769–74.

REPETTO, R. (1986). *Skimming the Water: Rent-seeking and the Performance of Public Irrigation Systems.* World Resources Institute, Washington, DC, USA: 46 pp.

REPETTO, R. (1987). Population, resources, environment: an uncertain future. *Population Bulletin,* **42**(2), Population Reference Bureau, Washington, DC, USA: 44 pp.

REPETTO, R. (1990). Deforestation in the tropics. *Scientific American,* **262**(4), pp. 36–9.

REPETTO, R. & GILLIS, M. (Eds) (1988). *Public Policies and the Misuse of Forest Resources.* Cambridge University Press, New York, NY, USA: xiii + 432 pp., illustr.

REPETTO, R., MAGRATH, W., WELLS, M., BEER, C. & ROSSINI, F. (1989). *Wasting Assets: Natural Resources in the National Income Accounts.* World Resources Institute, Washington, DC, USA: vi + 68 pp.

ROMM, J.J. (1992). *The Once and Future Superpower*. William Morrow & Co., New York, NY, USA: 320 pp.

ROMM, J.J. & LOVINS, A.B. (1992). Fueling a competitive economy. *Foreign Affairs*, **70**(5), pp. 46–62.

ROSENZWEIG, C. & PARRY, M.L. (1992). *Implications of Climate Change for International Agriculture: Global Food Trade and Vulnerable Regions*. Environmental Protection Agency, Washington, DC, USA: [not available for checking].

ROSENZWEIG, C., PARRY, M.L., FISCHER, G. & FOHBERG, K. (1993). *Climate Change and World Food Supplies*. Environmental Change Unit, University of Oxford, Oxford, England, UK: iii + 28 pp.

SADIK, N. (1991). *Meeting the Population Challenge*. United Nations Population Fund, New York, NY, USA: 48 pp.

SADIK, N. (1993). Women, population and development. Chapter 2 (pp. 37–70) in *Environmental Challenges [1]: From Stockholm to Rio and Beyond* (Eds M. NAZIM & N. POLUNIN). Energy and Environment Society of Pakistan & Foundation for Environmental Conservation, Geneva, Switzerland: vi + 284 pp., illustr.

SADIK, N. (1996). *State of the World's Population*. United Nations Population Fund, New York, NY, USA: [not available for checking].

SCHNEIDER, S.H. (1989). *Global Warming: Are We Entering the Greenhouse Century?* Sierra Club Books, San Francisco, California, USA: xiv + 317 pp.

SCHREIBER, G. & CLEAVER, K. (1992). *Sub-Saharan Africa: Population, Agriculture and Environment*. The World Bank, Washington, DC, USA: [not available for checking].

SENANAYAKE, P. (1995). Guest Comment: Reflections on the UN's International Conference on Population and Development. *Environmental Conservation*, **22**(1), pp. 4–5.

SHAW, R.P. (1989). Rapid population growth and environmental degradation: ultimate *versus* proximate factors. *Environmental Conservation*, **16**, pp. 199–208.

SHAW, R.P. (1992). The impacts of population growth on environment: the debate heats up. *Environmental Impact Assessment Review*, **11**, pp. 11–36.

SINDING, S.W. (1992). *Getting to Replacement: Bridging the Gap Between Individual Rights and the Demographic Goal*. The Rockefeller Foundation, New York, NY, USA: 16 pp.

SINDING, S.W. (1993). *The Demographic Transition in Kenya: A Portent for Africa?* The Rockefeller Foundation, New York, NY, USA: 15 pp.

SIVARD, R.L. (1991). *World Military and Social Expenditures 1991*. World Priorities, Washington, DC, USA: 64 pp., illustr.

STRONG, M.F. (1992). Secretary-General's Speech to UN Conference on Environment and Development, Rio de Janeiro, 4 June 1992. United Nations, Geneva, Switzerland: [not available for checking].

SUMMERS, L. (1992). The most influential investment. *Scientific American*, **267**, p. 132.

TOLBA, M.K., EL-KHOLY, O.A., EL-HINNAWI, F., HOLDGATE, M.W., MCMICHAEL, D.F. & MUNN, R.E. (Eds) (1992). *The World Environment 1972–1992: Two Decades of Challenge*. Chapman & Hall, London, England, UK: xi + 884 pp., illustr.

UNION OF CONCERNED SCIENTISTS (1992). *World Scientists' Warming to Humanity*. Cambridge, Massachusetts, USA: [not available for checking but quoted on p. 19 of Limited Geneva Edition preceding present update — *see* Preface to this volume].

UNITED NATIONS (1992). *Long-Range World Population Projections: Two Centuries of Population Growth, 1950–2150*. United Nations, New York, NY, USA: 35 pp.

UNITED NATIONS CONFERENCE ON ENVIRONMENT AND DEVELOPMENT (1992). *The Earth Summit's Agenda for Change: Agenda 21 and Other Rio Agreements*. United Nations, Geneva, Switzerland: [not available for checking, but *Agenda 21* said to be on pp. 21–4].

UNITED NATIONS DEVELOPMENT PROGRAMME (1991). *Human Development Report 1991*. Oxford University Press, New York, NY, USA: 205 pp., illustr.

UNITED NATIONS DEVELOPMENT PROGRAMME (1992). *Human Development Report 1992*. Oxford University Press, New York, NY, USA: 201 pp., illustr.

UNITED NATIONS DEVELOPMENT PROGRAMME (1993). *Human Development Report 1993*. Oxford University Press, New York, NY, USA: 216 pp., illustr.

UNITED NATIONS ENVIRONMENT PROGRAMME (1991). *Status of Desertification and Implementation of the UN Plan of Action to Combat Desertification*. United Nations Environment Programme, Nairobi, Kenya: [not available for checking].

UNITED NATIONS POPULATION FUND [UNFPA] (1992*a*). *The State of World Population 1992*. United Nations Population Fund, New York, NY, USA: 46 pp., illustr.

UNITED NATIONS POPULATION FUND [UNFPA] (1992*b*). *Population Growth and Economic Development: The Policy Response of Governments*. United Nations Population Fund, New York, NY, USA: [not available for checking].

UNITED NATIONS POPULATION FUND [UNFPA] (1992*c*). *Recent Developments in Research Into the Relationship Between Population Growth and Economic Development*. United Nations Population Fund, New York, NY, USA: iv + 64 pp.

UNITED NATIONS POPULATION FUND [UNFPA] (1993). *The State of World Population 1993*. United Nations Population Fund, New York, NY, USA: 54 pp., illustr.

US NATIONAL ACADEMY OF SCIENCES & [BRITISH] ROYAL SOCIETY (1992). *Population Growth, Resource Consumption, and a Sustainable World*. US National Academy of Sciences, Washington, DC, USA & Royal Society, London, England, UK: [not available for checking but quoted on p. 19 of Limited Geneva Edition preceding present update — *see* Preface to this volume].

WILSON, E.O. (1992). *The Diversity of Life*. The Belknap Press of Harvard University Press, Cambridge, Massachusetts, USA: 424 pp., illustr.

WORLD BANK (1990). *World Development Report 1990: Poverty*. Oxford University Press, New York, NY, USA: xii + 260 pp., illustr.

WORLD BANK (1991). *World Development Report 1991: The Challenge of Development*. Oxford University Press, New York, NY, USA: xii + 290 pp., illustr.

WORLD BANK (1992). *World Development Report 1992: Development and the Environment*. Oxford University Press, New York, NY, USA: xii + 308 pp.

WORLD BANK (1993). *World Development Report 1993: Investing in Health*. Oxford University Press, New York, NY, USA: xiii + 171 pp., illustr.

WORLD COMMISSION ON ENVIRONMENT AND DEVELOPMENT (1987). *Our Common Future*. Oxford University Press, New York, NY, USA: xv + 400 pp.

WORLD HEALTH ORGANIZATION (1992). *Reproductive Health: A Key to a Brighter Future*. World Health Organization, Geneva, Switzerland: xiii + 171 pp., illustr.

WORLD RESOURCES INSTITUTE (1992). *World Resources 1992–93*. Oxford University Press, New York, NY, USA: xiv + 385 pp., illustr.

ZAHR, C.A. & ROYSTON, E.A. (1993). *Global Factbook on Maternal Mortality*. World Health Organization, Geneva, Switzerland: [not available for checking].

3

Human Population Prospects

by

ROBERT ENGELMAN

Director, Population and Environment Program,
Population Action International,
1120 19th Street NW, Suite 550,
Washington, DC 20036,
USA.

INTRODUCTION

The emergence since World War II of authoritative demographic projections has brought to discussions of human population prospects an unwarranted sense of complacency. Because the projections are generally accepted as expert and reliable, non-demographic analysts tend to see projected population growth as an 'exogenous factor' — an inevitable, unstoppable force in human affairs. On most flow-charts of human and environmental interconnections, for example, population flows in from outside the system that is being studied — seemingly immune to the influence of anything within the system itself.

According to this view, human population will have specific magnitudes — potentially with specific impacts — at certain times in the future, all because the official projections have indicated so! Moreover these projections are widely considered to be reliable, because demographers employed by major organizations — such as the United Nations and the World Bank — make them in seemingly mysterious ways, and those demographers are not subsequently dismissed from their posts when they turn out to be wrong. A result is the common but misleading statement that the world's population 'is expected to double', or even that it 'will double again' in size by around the middle of the next century.

It is my intention in this Chapter to discuss human population prospects while challenging public perceptions of population projections, despite — or perhaps to some extent because of —having employed them frequently in research efforts. As a non-demographer, my appreciation of, and even admiration for, the courage and hard work of those who prepare projections is immense. Population projections are neither of necessity inaccurate nor useless, but important and even valuable tools for considering the human prospect. They are, however, liable to be misunderstood as reliable guides to the future of human numbers, and such misunderstanding has potentially hazardous consequences. In particular, the apparent mathematical precision of projections encourages the misconception that there is nothing anyone can *do* about population growth, when actually there is very much we can — and should — do. The usefulness of projections could be enhanced by much more open discussion of the real relevance of assumptions that underlie them, and an occasional challenge of some of those assumptions. The challenge presented here is based on certain key principles that deserve elaboration in advance for properly sceptical readers:

Firstly in discussing the prospects of almost anything, especially as complex as human population, humility is desirable and often important. As a meteorologist once said, prediction is hazardous, especially about the future: in trying to peer into the future, we confront a grand mystery with an infinite capacity for discontinuity and surprise. Nor should we ever pretend otherwise!

Secondly, prediction of human behaviour is necessarily subjective. Scientific culture shudders at the idea of subjectivity, yet science has produced no experimental technique that outperforms simple human experience in predicting how people may be expected to behave in particular circumstances or situations. The public has been lulled by a statistical mystique about demographers' work in projecting future population. The process is only objective in so far as it is made manageable by a handful of consistent assumptions. Among these is a lack of any surprises; for how can surprise be predicted, despite the fact that some is practically inevitable? Related to this is the assumption of a catastrophe-free future and an absence of interacting feedbacks among fertility, mortality, and migration, the three of them being key variables of demographic analysis. Because assessment of these assumptions is necessarily subjective, much that follows is based on the experience and judgment of this former science journalist and current population researcher for an organization that works to make voluntary family planning services universally available.

Thirdly, consideration of population prospects ideally should be an interdisciplinary endeavour that takes into account the many factors — economic, social, and environmental — that influence demographic variables. Debate on the Earth's human 'carrying capacity' has a history going back to the time of the Cambridge (England) don Thomas R. Malthus (*cf.* 1798), at least in terms of the number of people the planet's farmland and fisheries etc. could theoretically feed. This exercise continues to this day (FAO, 1984; Heilig, 1993; Smil, 1994), with results latterly giving increasing grounds for anxiety albeit not despair. There have been few efforts, however, to integrate into projections of fertility, mortality, and migration, any comprehensive information on the changing availability of the natural resources necessary for health and life (Lutz *et al.*, 1993). Even more difficult would be to apply, systematically, social and economic data to population projection. It would be rewarding, nonetheless, if some demographers tried taking some risks in this area to see what emerged.

Finally, in dealing with the future it is more useful to consider that which *could* be, rather than that which *will* be. The former category is so much larger in scope, so much closer to the grasp of current insight, and instills so much more hope for the future which our children and grandchildren will inherit, that it is puzzling to consider why the second category occupies the stature that it does. We have it in our power to influence significantly our demographic future. What follows will concern, above all, the population prospects that *we could claim for our species if we chose and took the necessary steps to do so.*

PAST AND PRESENT REALITIES

The future of world population strides forth from its past and present, being the cumulative product of thousands of generations of human reproduction. Much of the weakness in the current discussion about population's impacts stems from the fact that researchers limit their time-frame to the growth occurring in single years or decades, without contemplating the broad sweep of past and future time.

Demographers cannot speak with certainty about the history of population dynamics before the first censuses were held in the 18th century. Nonetheless, the historic evidence suggests that our species totalled no more than a few tens of millions of individuals throughout its prehistory of hunter–gatherers, and then rose into the low hundreds of millions after the dawn of agriculture. Thereafter, human population was relatively stable — perhaps occasionally declining, but most often growing very slowly at a small fraction of a percentage point annually — until about the time of the Renaissance in Europe (Livi-Bacci, 1992; Cohen, 1995a, 1995b).

In the two main centres of population at that time, Europe and China, gradually improving agriculture and better understanding of health and sanitation led to slightly higher population increases than had been experienced previously. The reason for this population increase was not any significant increase in the number of children

who were being born, but the decline in the proportion of people —
especially children and mothers — who were dying in any given
year. It is, in fact, decreasing death-rates far more than increasing
birth-rates that fuel most population growth. (The exceptions are
cases in which modern industrialized nations with stable and low
death-rates experience 'baby booms' or increases in net immigration, or both.)

As the Industrial Revolution advanced early in the 19th century,
human population surpassed 1 thousand million souls for the first
time in history, and growth in numbers accelerated even more.
Advances in agriculture and sanitation continued, while improvements in medicine and medical care contributed to the acceleration
of population growth — especially when once the principle of
immunization was understood and widely practised. The emergence
of mass-education undoubtedly helped, as well by disseminating
both knowledge and broader attitudes that enabled parents of modest
means to save more of their infants and children from dying, particularly by learning how to help them to survive.

Also important to population growth was the occupation by
Europeans of the then-thinly-populated Americas. This opened up
more than one-quarter of the world's land-masses to cultures of
which the technologies enabled them to bring four, six, or even eight
or more, children up to reproductive age for each woman. And, in
fact, some of humanity's highest rates of population growth —
approaching 3% annually — were experienced in the United States
during much of the 19th century. Considering the technological and
social systems of the 19th and early 20th centuries, Europe was a
crowded continent, and the opening-up of North America provided a
destination for the largest migration so far in human history.

Recent World Population Growth

World population continued to grow at varying rates around the
globe, passing two thousand millions by AD 1930 and approaching
2.5 thousand millions by the outbreak of World War II in 1939. The
end of that global war in 1945 marked a 'demographic watershed',
as education and knowledge of modern sanitation, immunization,
and farming techniques, spread practically throughout the world.
The result of this latest and most impressive victory over early death
is evident in every region and major city. The planet now sustains
more than **5.8 thousand million** human beings, nearly half of them
in urban areas. Nearly three out of every five persons live in Asia,
and more than one in three Asians is Chinese.

Each of the other major world regions is home to several hundred
million people, but the populations of each continent are growing at

different rates: Europe, with **729 million** people, is growing very slowly at a mere one-tenth of 1% annually. North America (mostly the United States, and Canada), with **302 million** people, is growing more rapidly at about 1.1% annually. Asia, with **3.54 thousand million** people, is growing at an annual rate of about 1.6%. The Latin American and Caribbean region, with about **492 millions**, is growing more rapidly still, at about 2% annually.

Standing apart from the rest of the world demographically is Africa, with **758 millions**, in most of which population growth has continued for decades at nearly 3% a year, with only some signs of falling. The average of all these uneven rates of growth world-wide is now slightly less than equivalent to that of Asia, being **just under 1.5%**.

This global growth-rate itself proclaims a signal development in world population dynamics: growth is slowing! The annual rate peaked at 2.1% in the late 1960s and has drifted down ever since. When a growth *rate* slows, however, *growth* itself continues until the rate reaches zero. And as the size of world population increases, more modest rates of growth can still add larger annual increments to the population-base. This has occurred: while the highest rates of population growth in the past saw only about 72 million people added to world population each year, the current lower rates of growth are adding about 81 million people annually, and this number has gone as high as 88 millions annually.

This increment has now begun to fall — a moment, indeed, in human population history — and may well continue to fall. In a world without surprises, the projections inform us, population growth is projected to be sufficiently modest that even with a larger population-base, the added numbers would become smaller each year, until eventually global births would equal deaths and the world's human population would stop growing.

Most of the easing of world population growth-rates occurred in the 1970s, as a response in part to the spread of organized family planning efforts in developing countries in that time-period. Fertility was also declining rapidly in industrialized countries, often falling for the first time in history below the approximately two-children-per-couple average that is necessary, in the absence of immigration, to replace each generation with the one that follows. The decline in fertility and population growth-rates apparently both decelerated in the 1980s — a demonstration of the fallibility of assumptions that fertility decline is inevitable and continuous. In the past few years, however, the decline in fertility appears to have resumed — and, indeed, to have finally begun in parts of sub-Saharan Africa.

The significance of this for the future of population is potentially large. Currently, throughout the developing world, women are seek-

ing to have smaller families than their mothers and even their older
sisters had, and they increasingly have the means to achieve the
family size which they seek.*

(The emphasis here is on women's child-bearing, disregarding
the role of men, because women are the primary practitioners of
contraception, and they alone bear children. Many women have the
support of their husbands in planning smaller families, but many do
not. Surveys indicate that, in most developing countries, increasing
numbers of women would like to have no more children [Holdgate,
1994] or anyway fewer children than their partners seek. This is not
surprising, for while both parents share in the joy and pride of
parenthood, women endure pregnancy and childbirth and then perform the vast majority of child-raising tasks. A major determinant of
future population size is the extent to which men support women's
ambitions to have fewer children. Another is the extent to which
women can put into effect their own decisions when the couple do
not agree.)

In much of the industrialized world, desired family-size seems
relatively stable. It may even be rising slightly in some countries;
but employment and housing conditions discourage many women
from having as many children as they would prefer under otherwise
ideal conditions. Where access to effective contraception and safe
abortion is good, women tend to have the childbirths they want,
given their life circumstances, and total fertility remains below the
'replacement level' of slightly more than two children per woman.
(The figure is slightly higher than an even 2 to allow for the fact
that, even today, some children do not survive to their own reproductive age.) This pattern is characteristic of most of Europe,
including formerly communist countries, and of Japan. It is also
evident in such developing countries as Hong Kong, Singapore,
South Korea, and Cuba.

Mortality and Migration also Condition Population

The two other variables that shape world population (aside from
fertility as we have defined it) are mortality and migration. Death-
rates, expressed as the number of deaths per thousand people in any
given year, continue to fall in most places around the world, which
means that life expectancy continues to rise. The dominant influences here are at both ends of the age spectrum: relatively fewer
children are dying in their first few years of life, and higher proportions of adults are surviving to old age. Both of these trends have

* Here we cannot help recalling suggestions which have been made that male
fertility may be vulnerable to environmental stresses related to industrialization (*see* also
page 60). — Ed.

come in response to improved nutrition and health-care technology and delivery.

One result of improved child survival, all else being equal, is faster present and future population-growth than would otherwise occur. (The evidence suggests that, eventually, fertility does decline, in response both to parents' greater confidence in small families and the contribution of contraception to child survival through the healthy spacing of childbirths at intervals of two years or longer*.) The ageing of a population, by contrast, contributes only to current rather than future population growth, as the elderly are past their child-bearing years.

Population ageing — an increase in the mean age of a population — is an almost inevitable result of slowing population growth, as new generations become relatively smaller and the proportion of elderly people increases. This fact prompts some commentators, worried about the social and economic implications of high pro-portions of retired people, to call for policies that favour population growth through subsidizing higher fertility or encouraging immigr-ation; but these commentators ignore the simple fact that population growth cannot continue for ever on a finite planet. Eventually, some population-ageing must simply be accepted as the inevitable price of population stability. Within a few generations after a population is stabilized, in any event, the process of population ageing would halt, as each generation and age-cohort will be approximately equal in size. For population growth results from the potentially numerous descendants of surviving children, who would never have been born if their forebears had died before reaching reproductive age.

Demographers assume that mortality decline will continue — indeed, the UN demographers contemplate an optimal life-expect-ancy of almost 85 years being achieved even in Africa — placing some further upward pressure on the pace of population growth. The pace of mortality decrease, however, could moderate world-wide as further improvements in health-care and nutrition become more difficult to achieve. Moreover, in some regions, mortality trends have recently reversed course and are actually on the rise, raising questions about the inevitability of mortality decline.

Thus in the European former Soviet states in recent years, mor-tality has climbed unexpectedly and dramatically since the dis-solution of the Soviet Union. Circulatory diseases have taken an increasing toll, attributed by some to extreme stress in these so-cieties in transition. Environmental devastation may also play a role

---

*In answer to our query the Author writes: 'No one is certain why this is, but it could be because the mother then has greater nutritional resources (especially in relation to lactation) and more time to devote to her toddler(s) as well as to the newborn.' — Ed.

in what appears to be the most rapid decline in life expectancy for
men in recorded history. Rising death-rates and falling birth-rates
have, over the past few years, led to actual population decline —
deaths exceeding births — in these countries, which demographers
had not predicted.

— and AIDS etc.

In some pockets of sub-Saharan Africa, the AIDS pandemic is
reversing past progress on decreasing mortality rates — a special
tragedy in that African life-expectancy has never approached that of
the rest of the world. Although demographers usually hesitate to
factor epidemics into their projections of mortality, the impact of
AIDS is so pervasive and resistant to intervention that recent UN
projections have introduced increased mortality into the calculations
for sub-Saharan Africa. In this region, AIDS is projected to diminish
— though not reverse — population growth.

Some epidemiologists anticipate similar impacts in the near
future in India, Thailand, and other nations of Asia, where the
human immunodeficiency virus is well established. AIDS stands as
a kind of 'wild card' in projections of global mortality rates, but the
reality is that many infectious diseases — Cholera, Malaria, Dengue
Fever, and Tuberculosis, among others — appear to be increasing
their reach and scope, relatively unimpeded by medical efforts to
contain them. When the growing resistance of some disease organisms and vectors to the drugs and chemicals used against them is
taken into account, along with the hazards of ongoing environmental
change, the future of world-wide mortality rates appears more
uncertain than ever. A signal weakness of most recent population
projections is the assumption of continued mortality decline well
into the 21st century — despite the strength of such recent adverse
trends.

Influence of Migration and Urbanization

As we have already seen, migration has a major influence on
population distribution and growth. An estimated 100 million people
— nearly 2% of world population — live outside of their countries
of birth (UNFPA,1993). An estimated 22% of this number reside in
the United States, contributing one-third of that country's population
growth-rate if the U.S.-born offspring of recent immigrants are not
counted. But the majority of immigrants — including most refugees,
meaning those who have left their country of origin ostensibly to
escape persecution — actually live in developing countries. International migration may ease global population-growth slightly, as
people who migrate from poor to wealthy nations tend to have fewer

children than those who remain behind. Many migrants today, however, move between developing countries, and their fertility impacts are less-clear.

Population projections generally operate on the assumption that mass-migration across international borders is a temporary phenomenon that will not continue much past the turn of the century. This assumption is puzzling. On the one hand, many governments in Europe and elsewhere appear determined to reduce or even eliminate immigration. On the other hand, the demographic, environmental, and economic, imbalances that fuel migration are almost certainly increasing. Information and capital move with increasing freedom across borders, and it seems a bit arbitrary to project into insignificance all international migration as a demographic factor.

Within countries, no such assumptions about migration's future hold. The movement of rural people to cities has slowed somewhat, but it is not anticipated to stop at any time soon. World population is projected to be more than half urban by about AD 2010, with rising proportions in cities thereafter. About three-quarters of the next century's population growth is assumed to take place in urban areas.

The effects of urbanization on population growth may be significant, as women in urban areas tend to have somewhat fewer children than those in rural ones. City life generally combines relative scarcities of dwelling space with an abundance of information about the wider world. There is also likely to be more independence in towns and cities from the 'extended families' and cultural tyrannies of the countryside, and more access to education and family planning services. All of these factors favour smaller families, and a largely urban world is more likely to embrace the small family ideal than did the rural world of the past. Much depends, however, on how successful urban societies of the future prove to be: will they be able to deliver the sanitation and health-care, including reproductive health and family planning services, that enable women to control their child-bearing and their children's survival?

PROJECTIONS AND THEIR PERILS

The challenge for demographers is to understand the complex and uneven trends in fertility, mortality, and migration, and to consider to what extent they are likely to continue into the future. It is a mammoth task to assemble and process the demographic data needed for sophisticated population projection, all of them based on imperfect censuses and vital statistics gathered by national governments of varying capacities around the world. The demographers who develop the projections deserve praise for a difficult job in

general well done. Yet given the public perceptions which they
engender, the projections themselves deserve more critical evaluation than they usually receive.

The major population projections are published by the United
Nations Population Division and, until recently, by the World Bank.
(The International Institute of Applied Systems Analysis in Laxen-
berg, Austria, has begun issuing projections as well.) The United
Nations offers a 'medium' population trajectory that would produce
a global population of about 9.4 thousand million people around the
middle of the 21st century, compared with 2.5 thousand millions in
1950 and 5.8 thousand millions in late 1996. The world's human
population would then grow fairly slowly, levelling off at around 11
thousand millions early in the 22nd century (United Nations Population Division, 1996).

The single projection offered by the World Bank (1993) closely
resembles the UN's medium projection (UN Population Division,
1992). The UN demographers, though not those at the World Bank,
issued two alternative 'variants' or projections, low and high, at least
suggesting that different population trajectories are possible. Recent
long-range global projections that extend to AD 2150 included a total
of seven projections. One, the so-called 'constant-fertility' projection, suggested that if global fertility rates remained at current
levels, world population in the middle of the 22nd century would
reach an unimaginable 694 thousand million people.

The main projections, by contrast, suggest a world population
reaching anywhere from 4.5 thousand millions to 28 thousand
millions in AD 2150 (UN Population Division, 1992). Even more
recent country-by-country projections on to AD 2050 (UN Population Division, 1996) produced significantly lower numbers for the
first half of the 21st century, the result of faster-than-anticipated
declines in fertility. These projections suggest an AD 2050 world
population between 7.7 thousand millions and 11.2 thousand
millions.

In practice, however, most journalists and analysts take the UN's
'medium variant', or middle trajectory, to be the most probable one,
and it is often expressed as the 'expected' population future. This
seems at first reasonable, as the UN's medium projection has
suggested similar sizes for world population, both in AD 2050 and at
ultimate stabilization, to those found in the World Bank's single-
variant projection (World Bank, 1993). Such a reaction is further
encouraged by the fact that in its own public-relations efforts, the
United Nations news media office has indeed called the medium
projection the 'most likely'. This is technically inaccurate, as the
exercise of producing variant population projections requires demographers to establish the outer limits, low and high, based on

assumptions about demographic variables. The medium projection is arrived at by roughly halving the difference in the assumptions that give rise to the low and high projections.

This hardly makes the medium projection 'most likely'. True, neither the high nor the low extreme could be properly considered 'likely' — they are extremes, after all — but neither is there any special centre of gravity midway between them. Each projection could be seen as equally possible, along with any point or trajectory in the spectrum of demographic outcomes described by the high and low extremes.

Another important factor lending credibility to the UN's medium projection is the fact that, over the past two decades, it is the medium projection that has turned out to describe best the actual demographic developments. In the 1970s, UN demographers were projecting a world population close to 6 thousand million people for the turn of the millennium, and with less than four years left to go, world population does in fact stand at more than 5.8 thousand millions, growing just rapidly enough to reach 6 thousand millions some time in 1999.

Yet it is also the case that recent global population dynamics have presented demographers with few surprises in this time-period, with the exception of the AIDS pandemic beginning in the early 1980s. Fertility rates have declined a bit more rapidly than expected, but so have mortality rates, practically balancing out the situation. Migration has exceeded demographers' expectations in the past two decades, altering the projections for some countries but not for the world as a whole.

A relative absence of demographic surprise, however, has not always been the rule. Until the 1950s, demographers most frequently underestimated population growth, and not because they expected fertility to fall faster than it did. In fact, fertility increased in industrialized countries after World War II — the so-called 'baby boom' — but no demographer foresaw the epochal development. The real reason for the failure of the early population projections was that their Authors missed the fact that mortality was falling with accelerating speed. The country-by-country triumphs of sanitation, clean water-supply, antibiotics, and vaccinations, were the surprises that the demographers missed.

One interpretation of this uneven record is that, as demographers have improved their skills, their record has improved. Future projections should thus run closer to the mark of actual population change. This is certainly a reasonable inference. But it may also be that populations evolve, as species are now believed to, in a process of punctuated equilibrium. Historically, population dynamics have sometimes followed a fairly predictable path, while at other times

they have been jolted by sudden changes in birth- or death-rates.
Although it may encourage complacency, the relative lack of surprises in the past few decades is not predictive. We can never
extrapolate an absence of surprise into the future. And there may be
good reasons for believing, subjectively, that surprise and dramatic
change are more likely tomorrow than they were yesterday.

Consider the common statement that the trajectory of population
growth is 'certain' over the next two or three decades. Most demographers would concur that world population realistically cannot
reach its maximum at a lower level than about 7.75 thousand
millions, and not at that level before about the year 2040. This
statement has much to recommend it, because population momentum is indeed a powerful force in near-term population growth.
Because that growth has been at unprecedentedly high levels for the
past 50 years, each generation has been more numerous than the last
in most countries. Even if women of child-bearing age immediately
began bearing no more than two children each, births would exceed
deaths for some time. Most deaths would be occurring among older
adults, whose generations were smaller than either the generations of
child-bearing couples or than those of recently-born children. Hence
population continues rising until relative balance is achieved
between the cohorts of the child-bearers and the elderly. At that
point, deaths would roughly equal births in number.

This reasoning is unassailable — in the absence of surprise. But,
strictly speaking, no population growth, not even tomorrow's, is
really *certain*. It is harder than it once was to imagine a nuclear
holocaust, but by no means impossible. Astronomers blithely inform
us that comets and asteroids have collided with the Earth, and could
do so again in our lifetime. Obviously, in the unlikely event that a
nuclear war breaks out, or a comet hits our planet, all demographic
bets would be off. It may seem a fine point, but words like 'inevitable' and 'certain' overstate the case.

More importantly, such language lulls observers into a conviction
that no action in the present can influence the demographic near-
future. Yet there is a continuum of possibility related not only to
catastrophe but to less dramatic changes in both mortality and fertility. Increases in mortality could result from environmental degradation, food insecurity, infectious disease, economic or social disruption, or a combination of any or all of these and yet other factors.
With much of the developing world exposed to television and computers in the span of a mere decade or two, a revolution in fertility
patterns cannot be ruled out either. The likelihood of such changes is
discounted in projections, perhaps reasonably; yet such assumptions
receive no discussion when the projection results are released to the
public.

DUBIOUS ASSUMPTIONS

The environmentalists' critique of the assumptions behind population projections is apt to begin with their failure to consider the likelihood of environmental limits to population growth. Some sceptics might suggest that the embarrassing record of the prediction of such limits (Malthus, 1798; Ehrlich, 1968) has wisely dissuaded demographers from a similar risk to their reputations. The more likely reason is that there is no scientific consensus about the extent or even existence of such limits. Some of the potential constraints on population growth are addressed in the forerunner of this volume by Nafis Sadik, Norman Myers, and others (in Polunin & Nazim, 1994). The debate on environmental constraints to population growth has been long, prolific, and occasionally even bitter and personal. Only a few points will be mentioned here.

When potential or supposed environmental threats are disaggregated and examined in isolation, as they are in most of the literature regarding limits, they can often be made to appear individually manageable. This is especially the case when humanity's historic capacity to innovate and adapt is taken into account. By these arguments — many of them focused on economic growth as a proxy for human and Nature's well-being — a significant warming of global climate will 'merely' require farmers to use a bit more ingenuity — and perhaps more water, fertilizer, and pesticides — to maintain or improve yield levels according to current trends or projections. Water shortages can be resolved through market pricing, desalination, interbasin transfers, and ever-more dams. The degradation of soil matters little in view of the vast potentials of hybridization and biotechnology to boost crop and livestock yields further. If some nations find it difficult to feed their populations, the latter can pursue their comparative economic advantages elsewhere or else import food from surplus producers. If individual species go extinct, it could be countered that no particular species is critical to economic growth.

The problem with these statements is that each presumes a specific environmental development occurring in the context of 'business as usual' in all other environmental (not to mention social and economic) contexts. But environmental trends tend to occur simultaneously and synergistically. They may operate cumulatively (*i.e.* add to each others' impacts) sufficiently to reach or surpass critical natural thresholds of sustainability, and individually or collectively threaten the survival of human societies. Moreover if, as such ecological economists as Herman Daly argue, economies are subsets of, and ultimately dependent on, ecosystems (Daly & Cobb, 1994), and if individual happiness and morale are influenced by the conditions of daily life (the weather, access to clean water and sanitation, the price and quality of food, as examples), then social and political stability may be affected by environmental developments as well. All of these can interact, so that the impact of the whole of environmental trends on human life and death could be far greater than the sum of individual factions.

So it is not merely a question of whether death-rates will rise or fall. Environmental trends could influence birth-rates as well, through increases in involuntary infertility and intentional decreases in child-bearing. This issue is rarely explored, but anecdotal evidence and logic suggest that such 'fertility feedbacks' could reduce birth-rates in some countries. Infertility appears to be a rising problem in sub-Saharan Africa and the United States, although its epidemiology remains uncertain. Recent scientific studies have concluded that sperm counts have declined significantly in recent decades throughout the world, possibly in response to environmental changes. Male animals exposed to certain chemicals resembling the hormone estrogen appear to develop female attributes. Some researchers have suggested that rising exposure of women farmers to agricultural chemicals could be influencing reproduction, lactation, and maternal and child health.

Equally plausibly, declining environmental quality and rising scarcity of critical natural resources could be influencing the decisions of couples and women about child-bearing. The concept of 'demographic transition' is often interpreted to mean that poverty is inevitably accompanied by high fertility while increasing economic development concomitantly produces fertility decline. But social scientists have recognized in recent years that this apparent relationship between economic development and fertility is far more complex and varied than used to be thought.

Recent evidence indicates that increases in the status of women and improvements in their capacity to regulate their child-bearing, are far more likely to lead to fertility decline than national economic growth alone (Robey *et al.*, 1993). Indeed, shortages of income and housing are considered likely contributors in recent fertility declines in countries as varied as Italy and Kenya.

Economic Aspects of Child-bearing

Could environmental factors perhaps play a similar role in the fertility calculus? Carl Haub, a demographer for the Population Reference Bureau in the United States, recently found, in a survey of women in Belarus, that lingering effects of the 1986 nuclear accident at Chernobyl in Ukraine were discouraging many women from having additional children in the latter former Soviet state. And a recent World Bank study of the population–environment nexus in sub-Saharan Africa found that desired family size in the region tended to correlate with the *per caput* availability of arable land: where farmland was abundant, desired family size was relatively large; but where farmland was scarce, women wanted fewer children (Cleaver & Schreiber, 1993, 1994). It may be that, at least in some regions, the long-standing assumption that rural couples inevitably want many children to work the land is weakening as there is progressively less land for them to work.

In my own travels in the developing world, community development workers have occasionally shared the view that scarcities of agricultural land, fuel-wood, and clean water, are prompting women to seek help in spacing childbirths and otherwise plan their families. This appears more likely to occur in those cases where women are learning techniques for conserving scarce natural resources. It is even possible that the economic assumptions about parents' valuation of opportunity costs related to child-bearing could explain some of this shift: when farmland and other natural resources are abundant, labour is a limiting factor in family well-being and large families are rational. But where natural resources are scarce, the need to manage and conserve them becomes paramount. Under such circumstances, women are less able to afford the demands on their time, health, and energy, that come from frequent child-bearing and constant attention to children. It may then make more sense to have only a few children whose health can be maintained and who can be educated in the complexities of resource management.

When such a calculation occurs in the context of access to good-quality family-planning services and enhanced opportunities for women in the formal economic sector — to own farms or launch businesses, for example — the effect on fertility could be very powerful indeed. Add the steady march of urbanization, which tends to increase awareness of the family-size norms of other cultures around the world, and it begins to seem possible that fertility decline could advance more rapidly than demographers have assumed.

Such reasoning, unfortunately, provides little guidance to demographers developing population projections. It does remind us, however, that these projections in effect 'take sides', however unintentionally, in the great scientific debate over the impacts of population growth on human well-being. The projections seem to support the view that there are no limits to population growth, simply because none are assumed in formulating the projections!

The issue of projected fertility decline presents another assumption that underlies most population projections and receives even less attention than those about declining mortality-rates. This involves the population's so-called total fertility rate, or TFR — the number of childbirths a woman would have in her lifetime if she experienced the child-bearing rates that are typical of the average woman of her age during each 5-years' period of her reproductive life. The main world population projections assume that each country will eventually reach a TFR slightly above, slightly below, or precisely at, two children per woman, but will then settle precisely at the selected figure and remain there indefinitely.

This assumption has its roots in history and mathematical logic. For most of human history, the effective number of children who survived to become parents themselves cannot have been many more than two per woman, or else population would not have grown so slowly for most of human history. Incredible though it now seems, families in which only two children survived to maturity must have been the general rule even in Africa and India, which had relatively stable populations for hundreds of generations before they were colonized by Europeans.

Need of Personal and Governmental Discipline

It is possible that traditional modes of contraception — especially prolonged breast-feeding and post-partum abstinence — resulted in significantly lower birth-rates. The dominant influence on what is called the 'net reproduction rate', however, was the much higher death-rates of the past. An African woman of the 8th century, for example, may on average have given birth to six live babies. But, if so, the chances of any one of them surviving to become a parent were only about one in three, and average life expectancy probably hovered in the late 'teens and early twenties. Seen this way, population programmes in developing countries do not so much impose upon their citizens the alien modern influence of artificial contraception, as weaken the alien modern influence of persistent above-replacement fertility, brought about as an unintended by-product of lower death-rates (Cleland, 1993).

Even more important for demographic projection is the mathematical logic which dictates that something very close to replacement fertility must be achieved again in the near future. The reasoning again is essentially ecological: exponential growth cannot continue indefinitely on a finite planet. In 1974 Ansley Coale calculated that, at its then-current rate of growth (since reduced somewhat), human population would occupy every square foot of land on Earth within seven centuries, and within 6,000 years 'the mass of humanity would form a sphere expanding at the speed of light' (Coale, 1974). Faced with the mathematics of exponential growth, demographers assume that current population growth-levels are a historical aberration, and that humanity will return to historical near-replacement levels within not too many generations. The dramatic fertility declines of recent decades further justify this assumption.

Replacement Fertility the Needed Norm

Logically, then, unless the planet Earth can support hundreds of thousands of millions of human beings, replacement fertility will need, sooner rather than later, to become again the norm. Actually, there is no guarantee that replacement fertility itself will always be two children per couple. If society failed to contain infant and child mortality-rates at their current historic lows, replacement fertility could rise. Already today, the replacement fertility-rate in high-mortality countries such as Ethiopia is as high as three children per couple. In past societies in which the life expectancy of women was as low as 20 years, replacement fertility would have been 6.5 children per woman. Obviously, no one would wish to envision such a future; so nowadays the conventional assumption is that replacement TFR is always only slightly higher than two.

Practically speaking, the developers of projections make their best guess as to when total fertility-rates will reach something close to the replacement level of just over 2 and stay at this level. The United Nation's most recent long-range low, medium, and high, variant projections are based on the assumption that total fertility-rates will stabilize, sometime before AD 2100, at about 2.05 (medium projection), 2.5 (high projection), or 1.7 (low projection).

The oft-cited 'medium projection' assumes that couples and women in industrialized countries will also settle for a TFR of slightly more than two children each. This seems reasonable — indeed, the populations of most developed countries have already achieved replacement fertility or something reasonably close to it. Most of these populations actually passed right through replacement fertility, however, and have settled, if they have settled at all, at a TFR *below* two children per couple. In many European countries and even such 'developing' states as Hong Kong, Singapore, and South Korea, fertility rates now sit at levels that will lead (or have already led) to population decline. The 'medium projection' assumes that women in these countries will eventually, in effect, 'come to their senses' and begin having the number of children needed to prevent depopulation of their national territories.

How realistic, however, is the assumption that any society will reach replacement fertility, from either above or below it, and then remain there? Is there something magical about this figure of an average of 2.05 children per woman? Is this a family size which men and women are likely to choose voluntarily — perhaps in recognition of the importance of eventual stabilization of the population in their countries and on the planet as a whole? Or is it the family size that governments will succeed in encouraging or inducing through population programmes?

The logical response to these questions is that replacement fertility is more a demographic concept than a force of reproductive gravity for men and women. Several industrialized countries that have experienced replacement fertility have then moved on, without noticeable disruption, either to reduce their fertility even further or to return to higher levels of fertility. In some less-developed countries, especially in Latin America, overall access to family planning and such other influences as the education of women, appear to be conducive to achievement of replacement fertility, yet fertility levels have remained somewhat above replacement. It is difficult to point to any country in which the total fertility-rate has stabilized at any low level for a long period, as projections assume fertility will do.

The truth is that medium projections are based on the convenient demographic calculation that replacement fertility-level is inevitable everywhere — that it will be universally accompanied by low mortality and thus remain at the precise average of just over two children per woman, and that it will become a permanent feature of human reproduction. Demographers have to guess likely dates for achieving this steady state, draw straight lines from today's fertility rates to the future ones, and then calculate the populations that would result, assuming constant or improving life expectancies.

TFRs Dynamic, Responsive to Life Circumstances

The demographic experience of the world to date, however, suggests that total fertility rates are dynamic and highly responsive to the life circumstances of women and couples. Although there are good reasons to expect fertility decline to continue where families are typically large, there is no particular reason to assume that fertility rates will settle at an average of 2.05 or 2.5 or 1.7 children per woman. It seems especially unsound to assume either that below-replacement fertility will inevitably lead to population decline (as the rates may well rise if housing or other economic conditions improve) or that it will return to, and stabilize at, replacement fertility. The implications of dynamic and condition-specific fertility for the future of population growth could be substantial.

What other factors might cast doubt on the projections? Another important one is the critical issue of the timing of childbirth. The projections assume no significant changes in the ages at which women and girls first give birth to a child. Nor do they assume that mothers will wait longer than formerly between pregnancies before giving birth to subsequent children. It is the nature of calculating TFR, which is based on the numbers of children born to women of similar ages, that age at first childbirth and birth-spacing are only indirect issues.

Yet these two factors influence birth and population growth-rates with impressive force. If women wait longer than formerly before their first childbirth, and longer than formerly between each subsequent one, they contribute to an attenuation of generations that reduces birth-rates and slows population growth. They do this even if they have ultimately just as many children as they would have had with no birth-delay or spacing. (In practice, however, women who begun child-bearing late and practice child-spacing, tend to have fewer children.) Moreover, the demographic impact of these practices is immediate. Delayed births weaken population momentum, which is the force that propels near-term population growth even after replacement fertility is reached, because 'tomorrow's parents are already here today.'

Yes, they are here today; but if tomorrow's parents not only produced small families but also had them late and through widely-spaced births, the effects on population growth even in the next decade or two would be surprisingly large. John Bongaarts, a demographer with the US-based Population Council, has calculated that if the mean age of child-bearing in developing countries were to rise gradually by five years between today and AD 2020, and if global fertility rates immediately reached replacement, the population of these countries would stabilize by AD 2100 at about 7.3 thousand millions. This is 1.2 thousand millions fewer than would be the case if replacement fertility began immediately in the absence of any change in child-bearing age (Bongaarts, 1994). It is also, intriguingly, much lower than was suggested by the UN's low variant projection, a 'best-case' population scenario.

Effecting delays in childbirths and longer intervals between pregnancies would hardly be achieved through intrusive or coercive population-control measures. Such developments would result, rather, from better educational opportunities and more access to paid employment. Also important would be help in improving sexual negotiating skills among adolescent girls* and better access to a range of birth-spacing contraceptives.

The reason for stressing later and more widely-spaced childbirths is not so much demographic as public-health-related. The evidence is overwhelming that more women and children survive pregnancy

---

* Such encouragement of 'help in improving sexual negotiating skills among adolescent girls' reminds us of one we heard screaming on the radio 'we must have sex', and similarly revolts older generations who were brought up to proper self-discipline with consequent abhorrence of modern laxity and self-indulgence. To them it is deplorable that there is rarely any mention, much less due consideration, of personal discipline or abstinence in such considerations. — Ed.

and the first few years of life when the mother is no longer a teenager and when births are spaced at least two years apart. In fact, an Indian researcher P.D. Sharma recently calculated that, if developing countries could achieve goals for child survival set at a 1990 UN 'summit' on children — with rates of infant and under-five mortality reduced one-third in each developing country and in no country having more than 70 deaths per 1,000 children — world population would stabilize at a level 1 thousand millions lower than that suggested by the low UN population projection (Sharma, 1994).

One reason for the already widespread slowdown in population growth is that parents would no longer have more children than they preferred for fear of losing the children already born and sufficient to support them in old-age. But an equally powerful reason was that *the contraceptive prevalence-rates needed to assure this level of child survival would, on their own, result in a much lower population growth-rate.*

Combining the conclusions of Bongaarts and Sharma leads to an intriguing idea. Here are two potential policy-goals the achievement of which could result in a population-growth trajectory well below that of the UN's low-variant projection. Yet they depend not upon fertility reduction but on improving the health and survival of women and children! What if government population policies aimed at both these goals — encouraging raising the mean age of childbirth among developing-world women by five years and reducing the infant and child mortality-rates by one-third? As the essence of both goals is delayed childbirth, no doubt their achievement would have overlapping effects. Yet it seems plausible that, by such means, world population could be made to stabilize at a population size much lower than the UN low projection suggests.

Are such goals realistic? Demographers will be appropriately sceptical. The point is simply that a significant shift away from today's patterns of early and comparatively closely-spaced births is feasible but not factored into even low-population projection. One could argue as well that meeting such goals is at least as easy to imagine as stabilizing the planet's atmosphere, expanding its renewable freshwater resources, or halting the destruction of rainforests.

PROSPECTS AND POSSIBILITIES

Demographers point to three 'certainties' — I would call them 'near-certainties' — in the future of human population growth: the first of these contentions is that considerable further growth will occur before the World's population stabilizes or reaches a peak. The second contention is that the vastly greater proportion of this growth will continue to occur in the least-developed or developing countries, while the third is that, as population growth continues to slow down, national populations even in the less-developed countries will age dramatically.

Beyond this, we are left with the well-quantified projections of the United Nations and one or two other organizations. It is much less clear that these are at all reliable guides to the prospects for world population. A Chinese saying suggests that 'if you do not change direction, you are likely to end up where you are headed.' The projections point out where human population is headed, but not necessarily where it will go. If current trends continue, and fertility falls steadily towards replacement levels while life expectancy rises to the optimum, then the range of expectations for the future of human population is probably about what the projections describe. Certainly it will be very difficult to stabilize population at a level below 7.7 thousand million people without either rising mortality-rates — which no moral society could willingly accept — or delays and reductions in child-bearing beyond what seems possible today.

In peering into the future, it is useful to consider population projections — all the variants and scenarios, not just the medium ones — as a statistically sound basis for what would be most likely to happen in a future *without significant surprises*. Then we should constantly remind ourselves that demographers have constructed a series of artificial alternatives in which all change is gradual and limited. These alternatives can teach us about our options in the present, but the future is most unlikely to unfold exactly as they describe. It is difficult to keep in equipoise this seeming contra-diction — the use of plausible scenarios for purposes of research and education while reminding ourselves and the public that no single scenario can be considered likely in all its details. However, that is what is needed if these predictions are to be regarded as 'sound'.

What, then, are the prospects for world population? It is here that experience, values, and subjective judgment, combine to form what must be a personal and individual view. Clearly, we must loosen the grip that the medium projections have on the limited attention of policymakers and the public. We need at least to bring to greater attention the range of growth suggested by the low and high pro-jections for the next century and beyond. And, despite its necessarily artificial quality, we should hold forward the low projection as a vision worth working towards: it is not a 'target' but a hoped-for by-product of aggressively pursued development initiatives that slow population growth while serving more immediate human needs.

Demographers are not convincing in arguing that the low pro-jection lies on the very borders of the impossible. In most instances

in which the projections rest on unrealistic assumptions — especially on optimal life-expectancy for all, and continued young
average ages of child-bearing — logic and some evidence argue for
adjustments that would result in lower rather than higher population
growth. Birth-rates could fall more quickly than the projections
suggest. Unexpected declines are emerging in India and sub-Saharan
Africa, and government population programmes have proved
popular even in heavily Catholic Peru and the Philippines. Death-
rates, unfortunately, may end up being higher than the projections
suggest. Both of these factors could combine to produce an earlier-
than-expected peak in population size.

Chances of Equitable Population Limitation

The above considerations put an advocate of early population
stabilization in a tenable though sensitive position. It is eminently
possible that population growth could decelerate for both com-
mendable and deplorable reasons, namely a simultaneous mix of
improved access to family planning and more decision-making
power in the hands of women, combined on the other hand with
some increases in unintended infertility and in death-rates that no
one could applaud. Indeed, approximately such a mix (with access to
abortion substituting for the availability of good contraception)
appears to be responsible for a reversal of population growth in
some former Soviet states. The only responsible position for advocates of population limitation, it seems to me, is to work to bring
down child and maternal mortality while continuing to support universal availability of reproductive health and family planning services and the greater capacity of women to use the latter effectively.

Is a global rise in death-rates likely? Only hubris could prompt
yet another prediction. What is striking about the state of the
environment, public health, and political discourse today, is not any
specific threat, but the overwhelming scale of human activities,
within an array of interacting human and natural forces, that could
tilt the balance towards a rise in mortality.

Global Threats need Heeding

Humanity today is now crossing a series of significant environmental thresholds at a time when even democratic societies seem
disinclined to take such threats seriously or even help those whose
well-being is most threatened. These threats include: the early stages
of what could be significant human-induced climate change; the
peaking of the global fish-catch; growing scarcity of renewable fresh
water; massive degradation of agricultural soils and alarming shrinkage of those that remain; the global re-emergence of infectious

diseases; and increasing resistance among microbes and pests of all kinds to drug and chemical controls.

Human beings are an innovative species — indeed the only one with capability of choosing and means of guiding its own future. But in today's free-market economies, innovation follows not so much human need as profitable opportunities. Will it be profitable to extend and improve the lives of the poor? And, if not, will governments or other benefactors pay for the innovations that will be needed to accomplish that goal? Predicting the future of human mortality may relate as much to judgements about human nature and political currents as to the vulnerability of the natural resources, ecosystems, and wider ecocomplexes, that support our own and other animals' societies.

For mathematical and ecological reasons, total fertility rates must eventually reach or decline below replacement levels. Yet falling life-expectancy could perversely raise replacement levels above two children per woman. Even on the optimistic assumption that replacement fertility-levels will not increase, however, a two-child average family does not seem implausible. This is especially the case when one recalls that a total fertility rate of two is compatible with the presence of three, or occasionally four or more, children in some families, bearing in mind also that rather many couples remain childless or have only a single child. Adoption, of course, is an obvious but under-emphasized option for those wanting large families. But all that is required demographically is that a significant proportion of people of reproductive age choose or are forced to have only one child or remain childless. A replacement-fertility society would not have to impose a two-child norm. Rather, it could free women and men from all child-bearing norms, other than that of respecting personal choice.

Already 44%* of all human beings live in a country in which total fertility rates are at replacement levels or lower. In rural areas, land and fresh water are becoming increasingly scarce, encouraging new thinking about the benefits of small families. The rising necessity and growing costs to parents of education, food, and clothing, and the onward march of urbanization, contribute to the same reexamination of the costs and benefits of large and unplanned families. This is especially the case as more and more people are exposed to the global information network, with its enticing visions of options and possibilities beyond that merely of raising a large family.

---

*In requesting some final updatings to his Chapter the Author wrote (*in litt.* 20 March 1997) 'I have just recalculated this figure [which is] based on the new UN estimates.' — Ed.

Added to these social factors is the growing commitment of countries, with some notable exceptions, to develop and implement population policies and to base these last on improved access to voluntary family planning, reproductive health services, and improved overall opportunities for women. The consensus reached at the 1994 International Conference on Population and Development (ICPD) in Cairo has not yet produced the needed shift of financial resources to population and human development efforts. But the Conference succeeded in establishing an international standard for the work ahead. As governments search for guidance in dealing with demographic pressures, the ICPD's Programme of Action offers a set of strategies that could have a dramatically slowing effect on population growth while producing immediate improvements in the lives of men and especially women (*see* Senanayake, 1994; also the opening chapter of this anthology, contributed by the Secretary-General of that important Conference).

Women's Status and Power Increasing

Perhaps the two most important influences on the future direction of population growth-rates will be the power that women gain over their own lives and the access they and their partners have to a range of choices for deciding when and how often to have children. There is no certainty regarding either of these questions while women remain subject to external control, abuse, and even violence, practically everywhere. Moreover, while the number of women of reproductive age grows by about 23 millions each year, an estimated 228 million women — one out of every six of reproductive age in the world — lack effective contraceptive protection. This is a minimum estimate, excluding women who are not pregnant or trying to become pregnant, and including those who are using traditional contraceptive methods with high failure-rates (Alan Guttmacher Institute, 1995).

Nevertheless, there is reason for optimism. Historical experience suggests that movements for human rights, when once they have been launched, move inexorably forward. True, there are apt to be setbacks; but social and political evolution has been relentlessly towards expansion of equal or at least fairer treatment under the law. This is true of slavery, racial segregation, criminal justice, women's suffrage, and even the treatment of animals (both domestic and wild). In the absence of catastrophe, it seems inevitable that women will expand their influence in economic, political, and social, spheres and likely (and overall beneficial) that their rights and capabilities will be more widely respected in the next century than in this one.

Advances are also under way in the personal regulation of child-bearing. Although there are problems of access to a variety of safe and effective contraceptives, family planning is not a concept that moves backwards. The past three decades have seen contraceptive prevalence grow from 10% to 55% of developing-world couples. This suggests a generality of satisfied clients, and exemplifies the functioning of a powerful and pervasive force that is likely to become more so in the complex and hazardous times that lie ahead.

The overwhelming but still widely unmet potential for family planning is a major challenge for governmental and non-governmental actors alike. The idea of family planning often has its own 'threshold effect': at a particular level of acceptance by pioneers, demand begins to spread rapidly and family norms are quickly transformed. This occurred most impressively in the 1970s in south-eastern Asia, but examples can be found in all regions. Some degree of family-size limitation can probably occur without major governmental and private efforts to expand family planning services, but the scope for this is limited. Historically, fertility declined already without modern contraception in France and other countries in Europe. But the time-periods involved for these earliest demographic transitions were long, and other factors may have been at work — in particular strong motivation and cooperation among married couples (relatively fewer births occurred outside of marriage) combined with some abortion*.

Reproductive Health-care and Other Factors

Another open question about the future of world population is the willingness of nations to invest the financial resources needed — about $17 thousand millions a year world-wide by the end of this decade, which is a relative pittance compared with military spending — to provide universally available reproductive health-care. This would include access to means of family planning for all who seek it, combined with maternal and child health-care and the preventive services aimed at sexually-transmitted diseases. If the resources were invested wisely, something roughly resembling the low projection of population growth could be achieved even with continuing declines in death-rates.

Three complicating factors deserve mention here, because changes in any or all of them could have substantial impacts on population growth. Indeed, the average newspaper reader could conclude

---

* and, we would expect, on the part of our 'more serious' ancestors, personal discipline and practice of self-abstinence, even if they may not be directly germane to this discussion. — Ed.

— inaccurately — that they are the most important issues in the population field. The three factors are migration, abortion, and China's population programme.

Dealing as it does especially with world population prospects, this chapter enjoys the luxury of largely side-stepping the question of the prospects for migration — especially international migration. Globally, the most significant prospect in this connection is for continued urbanization of the planet, which should contribute to future declines in fertility. More speculatively, as *per caput* availability of critical freedom and enjoyment of Nature's resources shrinks further in many countries, migratory pressures are likely to increase.

In much the same manner as international migration, abortion is among the 'hair-trigger' issues in the population field. Although the demographic implications of abortion are rarely discussed, they are significant. While some 190 million pregnancies occur world-wide each year, an estimated 51 millions of these end in abortion — 21 millions in countries where the procedure is illegal. Given that the world's population is growing by just over 80 millions each year, it is obvious that elimination of abortion — without a proportional decline in unintended pregnancy — would spur population growth by dramatically raising birth-rates. On the other hand, the proportion of births that are desired at the time they occur varies from an estimated 76% in sub-Saharan Africa to a mere 38% in Latin America (Alan Guttmacher Institute, 1995). Wider access to safe and legal abortion around the world would undoubtedly increase the already high proportion of unplanned pregnancies that do not result in births.*

It is impossible to predict the future availability of safe abortion. The prospects for greater use of drug-induced or non-surgical abortion appear bright in developed countries and some developing ones. But this option may not become widely available where governments and/or religious institutions are powerfully arrayed against almost all abortion. Some population analysts note the strong correlation between access to safe and legal abortion and low fertility around the world. (In every country except the strongly Catholic Republic of Ireland and Poland, wherever total fertility rates have dropped below replacement levels abortion is legal and generally available.) But activists against abortion rights understandably shun this reasoning.

A more persuasive argument for abortion rights than the need to limit population is their importance to the health and survival of

---

* It would also clearly reduce the numbers of unwanted children and benefit humanity's prospects in respect of our long-time impression that a large or anyway disproportionate number of the world's worst and most tyrannical dictators have been *unwanted children*. — Ed.

women, especially poor women. An estimated 585,000 women die each year from causes related to pregnancy and childbirth, and about 75,000 of these deaths are the result of unsafe abortions (UNICEF, 1996). It may well be that, over time, a number of factors will converge to liberalize abortion laws — greater sensitivity to women's health and to women's rights, greater sympathy for the many older women concerned about foetal abnormality who will only risk pregnancy if abortion is an option, and, eventually, perhaps even the demographic case that legal access to safe abortion is among the factors which contribute substantially to population stabilization.

None of these arguments, however, seems able to convince the many people who believe that abortion is morally equivalent to ending the life of a person. Their values and views are not likely to change, and the real question is how influential abortion opponents will become in each country. The safest bet is that the status of abortion will continue as today — muddled and contradictory. High levels of illegal and unsafe abortion may continue in place of access to effective contraception and — as no contraceptive is foolproof — safe abortion that societies can offer if they choose and are allowed to.

The Unique Case of Modern China

The experience of China would make a fascinating case-study of population policy amidst natural resources' constraints — the country is desperately short of renewable freshwater and arable land for its 1.2 thousand million inhabitants — were it not for its tragic impact on many of its own citizens and on the international population-policy debate. China has recently reiterated that it considers its one-child policy a 'necessity', given its shortages of arable land and fresh water and the continuing population growth resulting from its still-young population age-structure. While the extent of forced abortion and other forms of reproductive coercion is open to debate, there is no doubt that the programme relies on both social sanction and outright coercion — the heavy hand of local bureaucrats is certainly evident in China's low average fertility — for its implementation.

The high visibility of China's excesses raises difficult questions in the population field. Some advocates of population stabilization inside and outside of China defend the programme on the grounds that the balance between its natural resources and its population is so perilous. Despite the international consensus that population policies must be based on reproductive freedom — as affirmed in UN meetings ranging from the Earth Summit in Rio in 1992 to the conference on women in Beijing in 1995 — a case can be made that reproductive decisions can have major collective and community impacts, and some of them can be problematical. Ought there to be genuine population control, or at least in the phrase of Joel E. Cohen (1995*b*), 'a market in the externalities of childbearing'?

This question, of course, must be answered on ethical as well as practical grounds, and cannot be addressed adequately here. What can be said, however, is that much of China's dramatic decline in total fertility rates was achieved in the 1970s under a voluntary programme before the one-child policy was introduced. Had Mao Tse-Tung not blocked consideration of major family planning efforts prior to the 1970s, population growth might have slowed from an earlier point and the excesses of the 1990s might not have been regarded as necessary. Today it appears that China will reap the ill-will of the international community and many of its own citizens before it ultimately stabilizes its population. Over the long term, stabilization is more likely to occur — and endure — on the basis of voluntary child-bearing decisions rather than government coercion.

National Optimum Population Policies Needed

Population stabilization cannot be built upon the kind of short-term changes in fertility that coercive population-control programmes may produce temporarily but cannot, I believe, sustain. To help countries with rapidly-growing populations to stabilize theirs, programmes and policies will have to succeed not merely on time-scales of political terms of office but over generations. And to succeed at this they will have to be based upon popular consent and participation. Population policies and programmes can help to serve the demographic goals of a society, but only by serving primarily the private and felt needs of couples and individuals.

As a historically closed society and the last major communist power, China is a special case. There is not now, nor is there likely to be soon, the public feeling of urgency about population growth elsewhere in the world that would justify emulation of China's ex-cesses. The future is likely to see less, rather than more, 'population control' — meaning government attempts to bring population size directly to a target range — just as it is seeing less, rather than more, economic and political control.

Should governments nevertheless aim for an 'optimum' world or national population size? Many commentators have suggested that such numbers could be identified and perhaps even arrived at. But there is good reason for scepticism, the world being so complex. The figure would vary substantially — even if we had the needed data and understanding, which we do not — depending on the environmental issue or natural resource chosen for examination. More importantly, there is no population policy imaginable that would fully respect human rights of the individual, and thus be worthy of unqualified support while also taking us precisely to this hypothetical demographic state. While population dynamics do respond powerfully to governmental and private initiatives, the very idea of population *control* is fundamentally unworkable. As long as human freedom is paramount among our values, reproductive freedom should and will be highly valued. We can no more control population than we can, truly and sustainably, control people themselves.

It makes more sense to work for better understanding among all people of the linkages between population and *environmentally* sustainable development. Policies can then tolerate and even encourage the lowest fertility levels that will be consistent with the free and responsible decisions of women and men to have the childbirths they desire. This goal needs to be achieved in the future; but meanwhile solutions to still-threatening environmental problems should be sought among non-demographic contributing factors.

CONCLUSION

Environmentally, militarily, socially, and politically, we live in especially hazardous times, but it is difficult to identify any area of human interaction that would not be placed on safer ground if human population followed a slower rather than a faster path of growth. Although technology has its perils, there is much we should do to brake the acceleration of environmental degradation. There is also much that we can do to advance environmentally-benign technology and much more to reduce *per caput* rates of natural-resource consumption. Slowing population growth is environmentally compelling because it operates on the entire range of environmental problems, with benefits of 'impacts avoided' that multiply exponentially into the future. Moreover, slowing population growth is ethically acceptable because it is most sustainably accomplished not by controlling people's child-bearing but precisely the opposite: by allowing them the reproductive freedom and health which they demand.

Merely supplying safe and effective contraceptive choices to all who want them is not enough. But that step alone would quickly reduce the average fertility-rate in the less-developed countries (in which it is most needed) by 25% or more, bringing world population stabilization within reach, and exerting a positive influence on virtually every aspect of environmental and natural resource management. Additionally, there is very much more that societies can do to ease the demands for large families while improving the health and lives of parents and their children — especially those of women and girls.

Even the best and most creative of the thinkers on the future of environmental and natural-resource concerns tend to shy away from the contribution that sound population policies and programmes can make to solving environmental problems. Few environmental analysts keep up with developments in the population, family planning, and reproductive health, fields. The intellectual hegemony of the medium population projection discourages creative consideration of how population policies can help to build sustainable societies. And the uninspiring image of trying to pressure all the world's parents simply to 'stop at two' dies hard.

The irony is that the evidence is now strong that feasible and freedom-enhancing human development efforts, duly launched by governments and private organizations, could result in a path of world population future that would undercut even the United Nations' low projection which peaks at 7.75 thousand millions in AD 2040 (United Nations Population Division, 1996). We know where we want to go and essentially how to arrive there. What is most lacking is the political will to invest the needed human and financial resources. This is clearly where the work should lie now that the world is entering an era of demographic hope — which in turn can help to cure us of environmental despair.

REFERENCES

ALAN GUTTMACHER INSTITUTE (1995). *Hopes and Realities: Closing the Gap Between Women's Aspirations and Their Reproductive Experiences*. The Alan Guttmacher Institute, New York, NY, USA: 56 pp., illustr.

BONGAARTS, J. (1994). Population Policy Options in the Developing World. *Science*, **263**, pp. 771–6.

CLEAVER, K. & SCHREIBER, G. (1993). *The Population, Agriculture and Environment Nexus in Sub-Saharan Africa* (Revised edn). The World Bank, Washington, DC, USA: xix + 229 pp.

CLEAVER, K. & SCHREIBER, G. (1994). *Reversing the Spiral: The Population, Agriculture and Environment Nexus in Sub-Saharan Africa*. The World Bank, Washington, DC, USA: xii + 203 pp.

CLELAND, J. (1993). Equity, security and fertility: a reaction to Thomas. *Population Studies*, **47**, pp. 345–52.

COALE, A.J. (1974). The history of the human population. Pp. 15–25 in *Human Population*. Scientific American, New York, NY, USA: viii + 147 pp., illustr.

COHEN, J.E. (1995a). Population growth and Earth's human carrying capacity. *Science*, **269**, pp. 341–6.

COHEN, J.E. (1995b). *How Many People Can the Earth Support?* Norton, New York, NY, USA: x + 532 pp.

DALY, H.E. & COBB, J.B., JR (1994). *For the Common Good: Redirecting the Economy Toward Community, the Environment and a Sustainable Future* (Second edition, updated and expanded). Beacon Press, Boston, Massachusetts, USA: viii + 534 pp., illustr.

EHRLICH, P.R. (1968). *The Population Bomb* (Revised). Ballantine Books, New York, NY, USA: xiv + 201 pp.

FOOD AND AGRICULTURE ORGANIZATION (FAO) (1984). *Land, Food, and People*. Food and Agriculture Organization, Rome, Italy: vi + 96 pp., illustr.

HEILIG, G.K. (1993). How many people can be fed on earth? Pp. 207–61 in *The Future Population of the World: What Can We Assume Today?* (Ed. W. LUTZ). Earthscan, London, England, UK (for the International Institute for Applied Systems Analysis): xx + 484 pp.

HOLDGATE, SIR M.W. (1994). Hopes for the future. Pp. 221–46 in POLUNIN & NAZIM *q.v. see* especially Fig. 10 'Percentage of Women Wanting No More Children' showing marked increases in the 1980s over the 1970s in all of the 14 countries cited worldwide.

LIVI-BACCI, M. (CARL IPSEN, transl.) (1992). *A Concise History of World Population*. Blackwell, Cambridge, Massachusetts, USA: xvi + 220 pp.

LUTZ, W, PRINZ, C. & LANGGASNER, J. (1993). World population projections and possible ecological feedbacks. POPNET, International Institute for Applied Systems Analysis, **23**, p. 1.

MALTHUS, T.R. (1798). *An Essay on the Principle of Population as it Affects the Future Improvements of Society*. [Reprint 1926, Macmillan, London, England, UK: 396 + xxvii pp.]

POLUNIN, N. & NAZIM M. (Eds) (1994). *Environmental Challenges II: Population and Global Security*. (Limited Geneva Edition for United Nations Population Fund [UNFPA] *et al.*) The Foundation for Environmental Conservation, Geneva, Switzerland: xi + 285 pp., illustr.

ROBEY, B., RUTSTEIN, S. & MORRIS, L. (1993). The fertility decline in developing countries. *Scientific American*, **269**(6), pp. 60–7.

SENANAYAKE, P. (1994). Guest Comment: Reflections on the UN's International Conference on Population and Development. *Environmental Conservation*, **22**(1), pp. 4–5.

SHARMA, P.D. (1994). *Under Five Mortality, Fertility and Population Growth: A Behavioral Model*. United Nations Children's Fund, New York, USA: 15 pp.

SMIL, V. (1994). How many people can the Earth feed? *Population and Development Review*, **20**(2), pp. 255–92.

UNFPA — *see* UNITED NATIONS POPULATION FUND.

UNICEF (1996). *The Progress of Nations 1996*. UNICEF, New York, NY, USA: 54 pp.

UNITED NATIONS POPULATION DIVISION (1992). *Long-range World Population Projections: Two Centuries of Population Growth 1950–2150*. United Nations Population Division, New York, NY, USA: ix + 35 pp.

UNITED NATIONS POPULATION DIVISION (1996). *World Population Prospects: The 1996 Revision, Annex I: Demographic Indicators*. United Nations Population Division, New York, NY, USA: 255 pp., Tables.

UNITED NATIONS POPULATION FUND (UNFPA) (1993). *The Individual and the World: Population, Migration and Development in the 1990s, The State of World Population 1993*. United Nations, New York, NY, USA: iii + 54 pp., illustr.

WORLD BANK (1993). *The World Bank World Population Projections, 1992–1993 edition*. Johns Hopkins University Press, Baltimore, Maryland, USA: vii + 515 pp.

4

Where is the Time-bomb Ticking?

by

Sir Shridath Ramphal, LLM (London)
*1 The Sutherlands, 188 Sutherland Avenue,
London W9 1HR, England, UK;*

Co-Chairman, the Commission on Global Governance;
formerly *President of the World Conservation Union (IUCN)*
and *Secretary-General of the Commonwealth;*
earlier *Foreign Minister of Guyana*

INTRODUCTION

A scientist of Cambridge University was reported to have
suggested at a Conference (see *The Times*, 9 August 1993) that
Britain's population should be reduced to some 30 millions — about
half of its present size* — so that the British people could enjoy the
optimum quality of life, while preserving their environment and conserving their natural resources. The scientist was not advocating
anything drastic, let alone draconian, by way of immediate action;
he merely reckoned that if public policies continued to encourage
small families, and 'fertility' was maintained at its current level of
1.8 children per family on the average, the population would slim
down to 30 millions in 150 years.

I start with this reference to make the point that, while the
problem of rapid population-growth is one faced essentially in the
less-developed nations, the problems of environmental impact and
of pressure on resources have to be confronted in both less-
developed and more-developed countries; indeed, in some crucial
aspects, more especially in developed countries. The principal

---

* An even more drastic reduction of the Swiss population was suggested to us
more recently in a submission to *Environmental Conservation* by a Swiss academic
Author of note whose paper we did not publish but who maintained that, on its own,
Switzerland could only support a maximum of 1.5 million people whereas its current
resident population is about 7 millions. — Ed.

cause of human pressure on the environment is human consumption and profligacy, for population acts as a multiplier. The total impact can therefore be reduced by moderating human numbers or *per caput* consumption; or, optimally and most effectively, by doing both.

None of the industrialized countries is, so far as I am aware, contemplating any significant programme to cut *per caput* consumption or make their population smaller; but ecological compulsions, leaving aside considerations of global equity, may well require them to do so in the not-too-distant future. Instead of waiting for the leisurely, above-cited pain-free 'Cambridge' process of a slow decline in numbers to produce results, the so-called 'developed' countries may need to opt for more resolute action to reduce consumption. But I will return to this later.

World population has recently been increasing by 1.7% a year, which does not seem too alarming a figure by itself. It is when the percentage figure is translated into absolute numbers — into the ever-mounting increases in numbers which occur each year — that the enormous dimensions of present-day demographic change become clear and really rather shocking.

When the Reverend Thomas R. Malthus, another Cambridge don, was writing his pioneering *Essay on the Principle of Population* nearly 200 years ago (Malthus, 1798), the world's human population was edging towards its first thousand millions. It took just over 100 more years to reach the second thousand millions. The third was added in 30 years, the fourth in 15, and the fifth in only 12 years. From 1960 to 1987, a period of 27 years and within the lifetime of most of our children, world population increased by 2 thousand millions — the figure it had taken all the millennia from the emergence of *Homo sapiens* until 1930 to reach! From 1930 — in our own lifetime — the world's human population has trebled. We have latterly been adding almost a thousand millions every 10 years — approximately equivalent to an India each decade, or to a Mexico each year.

FUTURE PROSPECTS: GLOOM BUT NOT DOOM

We cannot write with any degree of certainty about future population growth except for the next 20 to 30 years. Beyond that, demographers can only offer projections, extrapolating from current trends and making assumptions about variable factors such as the pace of economic growth and the spread of family planning. They can only *speculate* about future technological advances or ecological constraints. Estimates of what the numbers are likely to be in the longer term are therefore imprecise at best. So demographers offer us high, medium, and low, variants, and we use their medium projection as a working basis.

The latest medium projection by UN demographers has the world's human population, of 5.8 thousand millions in 1996, touching 10 thousand millions by the middle of the next century, then continuing to grow gradually for another hundred years or so before levelling off at about 11.6 thousand millions. If human behaviour were to correspond closer to the high projection, the numbers would swell by another two to three thousand millions before peaking; but there are even higher projections that take us to 19 thousand millions by AD 2100!

At the International Conference on Population and Development, held in Cairo in September 1994 (*cf.* Senanayake, 1994), delegates were sufficiently concerned by the gravity of the situation to agree that the world should try to ensure that population is stabilized by AD 2050 — rather than a hundred years later as is now expected — and at a level lower than the 10 thousand millions projected at that time. Achieving such a target will call for strenuous efforts and intensive global cooperation.

Whatever the increase, as much as 95% of future growth in human numbers is fully expected to be in the less-developed countries, whose populations are growing at an average rate of 2.1% against an average of 0.5% in the developed countries. There is no room to doubt that the pressure exerted by populations expanding at the former pace would have the most serious environmental impacts.

Scientific advances have so far helped to frustrate Malthusian predictions of doom. Some technological optimists, who feel that our scientific ingenuity will continue to defeat ecological challenges, may still be around. But scientific opinion at the highest level no longer shares such blithe confidence.

In an unprecedented joint statement in 1992, the Royal Society in Britain and the National Academy of Sciences in the United States issued a warning in these terms: 'If current predictions of population growth prove accurate and patterns of human activity on the planet remain unchanged, science and technology may not be able to prevent either irreversible degradation of the environment or continued poverty for much of the world.' (US National Academy of Sciences & [British] Royal Society, 1992).

These fears do not reflect a sudden loss of confidence among leading scientists but are influenced by the present scale of population increase and the extent of pressure already exerted on the environment. The impact of people on their habitat is, of course, a complex phenomenon that is affected by several factors: geography and climate, the distribution of wealth, the availability, ownership, and fertility, of land, the level of income, the economic policies pursued, technology, agricultural practices — all, and more, have a bearing on the scale of population changes and their impacts.

'GREEN REVOLUTION' PROBLEMS IN POOR COUNTRIES

Generally speaking in poor countries, as human numbers rise there is increasing pressure on the land. Farming becomes more and more intensive, fallow periods between crops are shortened or eliminated, and more and more pesticide and fertilizer are applied. The result is early exhaustion of the soil and degradation of the land. Farming is also extended to more fragile areas — leading to soil erosion or desertification — or to forest land, when the result includes loss of trees, of shelter, and of biodiversity. Poor people, needing land to grow food for themselves and their families, widely cause even more deforestation than loggers seeking timber!

Increase in yields per hectare may be possible in some countries, but 'Green Revolution' technologies have faced not just social costs but other problems as well. Soil erosion, salinization, and other forms of land degradation and ecological impacts, erase about half the gains from technology-based improvements in grain production, forcing and holding output gains below the rate of population growth.*

'ENVIRONMENTAL REFUGEES' INCREASING

Population increase exacerbates other environmental problems as well: water scarcity, inadequate sanitation, and lack of fuel-wood, are all parts of the chronic environmental experience of millions in the South. Rapid population increase also spurs migration from rural areas, adding to congestion in urban areas and the environmental and other strains that go with it. Migration has long been a survival strategy among the poor: it is no new phenomenon, but is now assuming worryingly new proportions. The term 'environmental refugee' has joined 'political refugee' in the humanitarian lexicon.†

Environmental refugees have so far been largely the products of desertification and famine in Africa, and the host countries which have had to accommodate and support them have also been in that

---

*As indicated with due documentation by Dr Norman Myers in Chapter 2. — Ed.

† *See*, for example, Chapter 9 below; also Westing (1992). — Ed.

region. Future movements, however, will not be confined to those causes or that continent.

Transborder movements apart, the likely future pressures on the cities of the Third World are sufficient cause for concern. In 1960, of the ten largest cities in the world, seven were in developed countries and only three in developing countries. Now the position has been reversed, as seven of the largest 10 cities are in developing countries. By the end of this decade there are expected to be 20 cities with more than 11 million people each, with 17 of these 20 in developing countries. Mexico City, the largest, may soon have as many people in it as the entire populations of The Netherlands and Belgium combined.

The trend to megacities, which *The Economist* has termed 'monsteropolises', is continuing unabated (*The Economist*, 30 May 1992). Today, 45% of the world's people live in urban areas; this is expected to rise to 65% by AD 2025. As situations in Calcutta and São Paulo testify, developing countries are already hard put to maintain even the barest environmental standards in their most congested urban agglomerations.

Neither rural nor urban areas of the Third World will, therefore, escape environmental degradation if population growth continues unchecked. The people of the less-developed world — particularly the poorest people — will be the principal victims. Whether from the lack of land or its diminished fertility, the shortage of fuel-wood, bad sanitation, or poor housing, and whether it is from hunger or malnutrition or water-borne diseases, it is the poorest people who will suffer most. It is therefore in the interests of those countries and their people that the problem of rapid population growth should be tackled resolutely.

FERTILITY RATE REDUCTION

Already countries in the Third World have begun tackling those problems of overpopulation with considerable success. Thus the overall number of children per woman (the 'fertility rate') has declined in developing countries (from 6.1 per woman in 1965–70 to 3.9 per woman in 1985–90). The result was some 70 millions fewer-than-projected births per year in the developing world throughout the later years of the 1980s — a momentous demographic change, given all the difficulties which developing countries have experienced. They achieved that decline from 6.1 per woman to 3.9 per woman in less than 30 years, whereas the time which it took to progress from a family size of 6.5 to 3.5 for the

United States was 58 years. For Indonesia (by no means the prime example of fertility decrease), it was 28 years.

In three decades the prevalence of contraceptive use has risen from 10% to over 50% in developing countries, and more and more women want to have no more children (Holdgate, 1994). The challenge ahead, for developed and developing countries alike, is to encourage and assist in the acceleration of these overall trends. It has been estimated that 100 million more women would use contraceptive services if they were available and of high quality (Catley-Carlson, 1993). If these needs were met, contraceptive prevalence could be raised to about 60%. But this is not the only need: let us remember that the demographic transition in Europe was achieved without the benefit of modern contraceptives.*

Economic development and rising prosperity have been primary bases of achieving fertility-decline in developed countries. If the world is serious about bringing down population growth-rates, our global community must help to create conditions in which couples in developing countries will consciously and voluntarily seek a lower number of children than their forbears. Reduction of poverty, improved standards of health and education, raising of the status of women and increasing their opportunities: these are the conditions that can increase the impact of available contraceptive services, as was broadly recognized at the 1994 Cairo Conference on Population and Development.

If we are to influence reproductive behaviour in poor countries in favour of small families, development — especially social development — is a necessary complement to effective family planning services. But this is precisely the time when development cooperation is weakening.

BRETTON WOODS REFORM

At a meeting in September 1993 of the North/South Roundtable, held in the very same place in Bretton Woods, New Hampshire, where the post-war economic order had been established nearly 50 years earlier, a call was made for a new framework for development cooperation (North/South Roundtable, 1993). It came with the recognition that one of the principal motivations for developmental assistance in the last few decades was to seek allies in the Cold War, and that, whereas the Cold War had ended, current allocations of

---

* Also that disciplined self-restraint can, and surely should, be an important factor.
— Ed.

assistance still carried its scars, retaining only slender links with the objectives of poverty alleviation and human development. The point was illustrated with examples that are pertinent to any discussion which looks to the centrality of development, and development assistance, in securing global sustainability. They were these:

- Egypt receives US $370 (or equivalent) of development assistance per poor person, compared with $4 for India.
- El Salvador received more US assistance in 1992 than Bangladesh, even though Bangladesh has 24 times the population of El Salvador and its citizens have only one-fifth of the income *per caput* of El Salvadorans.
- Only one-quarter of Official Development Assistance (ODA) goes to the ten countries containing three-quarters of the world's poor.
- Twice as much ODA *per caput* goes to high military spenders as to more moderate military spenders.
- About 95% of the $15 thousand millions in technical assistance every year is spent on foreign experts and equipment from industrial countries — rather than on national capacity-building in developing countries.
- Only 6.5% of bilateral assistance is earmarked for human priority concerns of primary health-care, basic education, safe drinking-water, nutrition programmes, and provision of family planning services.

Of even more direct significance in the context of the need to moderate population pressure, is the fact that only 1% of all development-assistance flows is devoted to population programmes.

COOPERATION BEYOND AID

The weakening of development cooperation 'beyond aid' is in my view even more inexcusable. The European Community's economic relations with Africa, the Caribbean, and the Pacific, are covered by the Lome Convention, which is a Convention for preferential market access; yet the substantial experience is that of continued trade-barriers (especially in products covered by Europe's Common Agricultural Policy such as sugar, beef, and vegetables, which are all still subject to quotas and market restrictions). Protectionism over a wide area is a fact of economic life for developing countries.

Despite all the talk about free trade and the insistence on developing countries opening up their markets, the United States and the EEC in particular have adopted a kind of '*apartheid* double-speak' in their range of right-sounding trade wrongs, *e.g.* 'orderly marketing arrangements' and 'voluntary export restrictions'. There is now a second generation of these perverse labels: 'managed trade', 'results-oriented negotiations', 'European preference', 'social dumping', and 'strategic industries'. So how does the Third World export, how does it sell, and, especially, how does it develop in ways which encourage the demographic transition?

When he was UN Secretary-General, Dr Boutros Boutros-Ghali gave a stark description of Africa's debt crisis: 'External debt', he said, 'is a millstone around the neck of Africa, ... easing the Continent's debt-burden must be a priority for the international community' (Boutros-Ghali, 1993 p. 12). Latin America is often thought of as the centre of the debt crisis. It is, if you are a Western bank and a creditor. But in fact Africa is considerably more 'debt distressed' than Latin America, where external debt is 37% of GNP, while for Africa it is over 100%. As the excellent OXFAM report 'Africa Make or Break' vividly described it: 'Every Zambian man, woman, or child, has the dubious privilege of owing the country's creditors around $766 — more than twice the average annual income level. Mozambique is embarking on a process of post-war reconstruction shackled by a foreign debt equal to four times its national income' (Oxfam, 1993). How does Africa make the transition to small families that economic transformation encourages?

This matter of financial resources for development has acquired a particular focus in relation to implementation of UNCED's *Agenda 21* (United Nations Conference on Environment & Development, 1992). At the Earth Summit there was no definitive conclusion, no firm commitment of 'new and additional resources'. The question of resources from rich countries for implementation of *Agenda 21* — the 20% of the resources required to catalyse the much greater contribution from the poor themselves — was deferred: at best, for consideration by the rich among themselves, or, at worst, *ad infinitum*. *Agenda 21* was agreed to, but in effect was made subject to the provision of resources needed for its implementation, the possibility of which becomes more and more remote as time goes on.

The logotype of UNCED depicted the Earth 'In Our Hands'. It asserted that 'sustainable development' required a shared effort by all the world's people — a partnership for survival in which each country has a role that is related to, and often integrated into, the roles of others. The partnership, of course, is not between equals. Developed and developing countries are unequal in responsibility for allowing matters to go wrong and in their capacity for setting things right. Aristotle, in his 'Ethics', advised us a long time ago that equity between unequals requires not 'reciprocity' but 'proportionality'. His dictum holds in this ultimate domain of environmental restoration. *Proportionality* must be the ethical touchstone of the role of developed and developing countries in their partnership for global survival through environmentally sustainable development.

CONSUMPTION EXPLOSION UNSUSTAINABLE

Why are we concerned about a human population explosion? Why do we think of it not as a flowering of our species but in the negative sense of an overgrown garden? If we are, as some of us believe, the best thing that has happened on Planet Earth, why shouldn't more of us be ever-welcome? There is good reason for us to worry, although we seldom present the case fully. The real and ultimate reason for our concern is sustainability – the sustainability of our life on the Planet. In scientific terms, it is described as Earth's 'carrying capacity'; less formally, it is our impact on The Biosphere when properly measured by what we use and what we waste.

When we ask whether Planet Earth can sustain double its present human population, the answer has to do with *consumption*. If we continue to draw from Nature at the rate which we do today — if, overall, we consume at today's level and continue with 'business as usual' — such abandonment to profligacy can scarcely be sustainable. The population explosion could then threaten our own and Nature's survival on Earth. Remember the words of the British and American scientists cited above, and the even stronger statement signed by 1,680 scientific leaders in 70 countries, including 104 Nobel Laureates, from which we quote as follows:

> 'Human beings and the natural world are on a collision course. ... The Earth's ability to provide for growing numbers of people is finite. ... and we are fast approaching many of the Earth's limits. Pressures resulting from unrestrained population growth put demands on the natural world that can overwhelm any efforts to achieve a sustainable future. ... No more than one [to] a few decades remain before the chance to avert the threats we now confront will be lost and the prospect for humanity immeasurably diminished.' (Union of Concerned Scientists, 1992).

Those leading British and American scientific bodies were concerned about consumption before stating that 'The future of our planet is in the balance', while the quoted Concerned Scientists concluded that 'The Earth's ability to provide for growing numbers of people is finite', so that 'the prospect for humanity [is becoming] immeasurably diminished.'*

---

*Here we cannot refrain from quoting from our Foundation's notepaper that, with human population as with so much else, 'small is sustainable'. — Ed.

In 1986 was published, by Peter Vitousek and others including the Ehrlichs, a study entitled 'Human Appropriation of the Products of Photosynthesis' (Vitousek *et al.*, 1986). Photosynthesis, of course, is Nature's solar-powered basic food-making process. Its production of, primarily, carbohydrates is the material that sustains all life. Only 70 years ago, human requirements took about 10% of the life-sustaining product of photosynthesis on land — 10% to sustain our annual growth — hence the prevailing belief, then and for some time afterwards, that Nature was limitless and inexhaustible. The Authors of that study had calculated that, by the mid-1980s, the 5 thousand million people then expected on Earth would have raised that appropriation to 40%. If our numbers double to the predicted more than 10 thousand millions, they concluded, it might well be impossible for human appropriation to double to 80% of the products of the world's land-based photosynthesis. In other words, it may not be possible — let alone desirable — for us to continue much longer on our present consumption-path.

To succeed in doing so, we would have to preempt in the process other animal life in a desperate human scramble to enlarge land cultivation and annexe all its product for ourselves. Science and technology may increase the solar-powered productivity of Nature's plants, but we are also cutting back production as desertification, urban growth, highways and runways, soil erosion, and pollution, all steadily decrease the extent of Earth's green cover and likewise its food-building potentiality.

In any event, after taking anything like that suggested 80% of Earth's life-sustaining material around AD 2050, Humankind would not be far from the absolute limit of 100%. Like a plague of locusts, we would have eaten ourselves 'out of house and home'. Like locusts we would move on; but the only thing left for us would be aquatic plant life, which our processing technologies would turn to human use, though most likely they would succeed in providing no more than fractional relief. Of course, well before our continental food-stores ran out, men, women, and children, would have become embroiled in a primitive internecine struggle for survival — a struggle over consumption and the materials to maintain it which would pale into tiny proportions the civil and other wars and local conflicts that already rage in so many parts of our increasingly overpopulated world.

INEQUITABLE CONSUMPTION PATTERNS

The problem, however, is not only the level of human consumption but also its asymmetric pattern. At present, about one-quarter of the world's population (mostly in industrial countries) accounts for about three-quarters of the world's net annual consumption of resources of all kinds. Thus the industrial world consumes 75% of the world's commercial energy, 90% of its traded hardwood, 81% of its paper, 80% of its iron and steel, 70% of its milk and meat, and 60% of its fertilizers. All together, the other three-quarters of the world's people must get by on the remaining one-quarter of the resource-pie.

Estimates vary as to the relative consumption of rich and poor in the world. The celebrated ecologist–writer Norman Myers has written, 'The average British family comprises two children, but when we factor in resource consumption and pollution impacts, and then compare the British life-style with the global average, the "real world" size of a British family is more like 10–15 children.' (Myers, 1993 p. 209). In UNDP's 1993 Paul Hoffman Lecture, the President of the Population Council, Margaret Catley-Carlson (1993 p. 3), said she reckoned that 'every child born in the North consumes over a lifetime from 20 to 30 times the resources, and accounts for 20 to 30 times the waste — year in and year out — of their counterparts in developing countries.' On the basis of the lower end of this estimate, the 1.2 thousand million people of the North could be taken as the equivalent, in relation to consumption, of 24 thousand million people in the developing world — which means that the people of the North already impose on the Planet the burden of a century of the unborn of the South.

As we have seen, the world's population has latterly increased at a rate of nearly 1 thousand million persons per decade, and is predicted by some to continue to do so for several decades more. Of those increases, only about 5% will be in the developed world. But from a consumption standpoint, because of the much higher average *per caput* use of resources in developed countries, that 5% added in the North will impose on the Planet a far greater burden than the 95% born in the South. *So where is the time-bomb ticking?*

The truth is that there are many explosions in the making. The 95% of world population growth that is projected to take place in the South is one of them — in sheer numbers alone, and with dire implications for people in the South. But the 5% of world population growth attributable to the North is a large or yet larger explosion — not in relation to numbers, but to what the numbers imply for the Planet. Perhaps the whole analogy of time-bombs ticking away is misleading; for the bombs have been detonated already — the explosions have occurred. What is needed is containment of them: rolling back consumption levels in the North and reducing population growth in the South.*

---

* *See also* Chapter 6 of this anthology. — Ed.

OBLIGATIONS OF NORTH AND SOUTH

There is a further point about consumption. It is palpable — or should be — that we cannot realistically project population growth in the South into the 21st century at the same levels of consumption that prevail there today. People in the South will not all remain abysmally poor. Already East Asia is showing the way to other developing regions, and not just in marginal development. By the first decade of the 21st century, the world's economic centre-of-gravity may have shifted away from North America and Europe to East Asia, perhaps to an East Asian Economic Community in which Japan may occupy a prominent place. Latin America may be next and then South Asia and, last of all, Africa — though assuredly not for many decades to come. Moreover, it is going to be in the world's interest — very specifically in the interest of reducing population growth-rates — that development should be a reality throughout the South.

But what development? Development which, through 'CNN' and 'Sky' and 'Star' and all the other communication wonders that lie in store for us, mimics the consumption culture of the West, requiring the Governments of the South to deliver a development which more and more approximates to the life-styles of the West? If so, we have to think of the burden on the Planet, in what the economists call 'real terms', not as a doubling but suffering something nearer a ten-fold increase in the world's 'effective' population — in short, of an Armageddon long past the horrors of the Vitousek *et al.* (1986) projection.

Our intelligence should tell us that we have to lower the levels of consumption in the North, that we have to lower the levels of population growth in the South, and that these are not an either/or proposition but both urgent imperatives.* Thus we have to concern ourselves, and persuade our ministers and policymakers to concern themselves, not only about population and development, but also about consumption as well; or at the very least we have all to convey the message that almost nothing we do about population and development is really going to suffice unless we do something earnestly and quickly about consumption also, starting immediately with *drastic reduction of wastage of all kinds.*

In the final analysis, the answer to the population problem lies in effective development: more real and environmentally sustainable development bringing improved quality of life in the South, and improved quality of development with lower levels of consumption in the North.

---

* *Cf. Our Global Neighbourhood: The Report of the Commission on Global Governance.* Oxford University Press, Walton Street, Oxford, England, UK: xx + 410 pp., 1995. — Ed.

REFERENCES

BOUTROS-GHALI, B. (1993). *Africa Make or Break: Action for Recovery.* Oxfam, *q.v.*, pp. 11 & 12.

CATLEY-CARLSON, M. (1993). *Explosions, Eclipses and Escapes: Charting a Course on Global Population Issues.* The 1993 Paul Hoffman Lecture, United Nations Development Programme, New York, NY, USA, 7 June 1993: 17 pp., typescript.

Economist, The (1992). [Issue of 30 May 1992], page 11.

HOLDGATE, SIR M.W. (1994). Hopes for the future. Pp. 221–46 in POLUNIN, N. & NAZIM, M. *Population and Global Security* (Environmental Challenges II). Limited Geneva Edition for UNFPA *et al.*, edited by NICHOLAS POLUNIN & MOHAMMAD NAZIM, published by the Foundation for Environmental Conservation, Geneva, Switzerland: xi + 285, illustr.

MALTHUS, T. R. (1798). *An Essay on the Principle of Population as it Affects the Future Improvement of Society.* Revised 1803; reprinted 1976, Norton, New York, NY, USA: [not available for checking].

MYERS, Norman (1993). Population, environment, and development. *Environmental Conservation,* **20**(3), pp. 205–16.

NORTH/SOUTH ROUNDTABLE (1993). *The United Nations and the Bretton Woods Institutions: New Challenges for the 21st Century.* Report of the North–South Roundtable Session, Bretton Woods, New Hampshire, USA, 1–3 September 1993, North–South Roundtable, 3 UN Plaza, New York, NY 10017, USA: 24 pp.

OXFAM (1993). *Africa Make or Break: Action for Recovery.* Oxfam, 274 Banbury Road, Oxford, England, UK: 38 pp.

SENANAYAKE, P. (1994). Reflections on the UN's International Conference on Population and Development. *Environmental Conservation,* **22**(1), pp. 4–5.

Times, The (London) (1993). [Issue of 9 August], page 3.

UNION OF CONCERNED SCIENTISTS (1992). *World Scientists' Warning to Humanity,* Cambridge, Massachusetts, USA: [not available for checking].

UNITED NATIONS CONFERENCE ON ENVIRONMENT & DEVELOPMENT (cited as UNCED) (1992). *Agenda 21* [not available for checking but for detailed treatment *see Agenda 21: Earth's Action Plan, Annotated.* (IUCN Environmental Policy and Law Paper Nr 27.) Oceana Publications Inc., New York, London & Rome: 683 pp.]

US NATIONAL ACADEMY OF SCIENCES & [BRITISH] ROYAL SOCIETY (1992). *Population Growth, Resources Consumption, and a Sustainable World.* US National Academy of Sciences, Washington, DC, USA & Royal Society, London, England, UK: [not available for checking].

VITOUSEK, P.M., EHRLICH, P.R., EHRLICH, A.H. & MATSON, P.A. (1986). Human appropriation of the products of photosynthesis. *BioScience,* **36**(6), pp. 368–73, illustr.

WESTING, A. H. (1992). Environmental refugees: a growing category of displaced persons. *Environmental Conservation,* **19**(3), pp. 201–7.

5

Population, Economic Change, and Environmental Security*

by

CLEMENT A. TISDELL, PhD (ANU)

*Professor & Head, Department of Economics,
The University of Queensland,
Brisbane 4072,
Australia*

INTRODUCTION

Population levels and their geographical distribution are key considerations in economic development and major influences on the state of the global environment. Virtually all economic development models include human population as a key variable, even though different theorists draw different conclusions about the implications of population increase for 'sustainable development'. Furthermore, diverse theories have evolved about socio-economic influences on population increases and the relationship of population levels with the natural environment. Such variety is to be welcomed, because it allows us to view the issues involved from varied perspectives and also reflects an expanding body of thought to meet changing circumstances. We are now better equipped than ever before to understand how socio-economic factors influence population increase, and how population increase is related to, and impacts on, the prospects for sustained economic development and for environmental security.

Population, environmental variables, economic well-being, and development, are closely interwoven (Tisdell, 1993c). For example, poverty may arise from an unproductive and unfavourable natural environment from a human viewpoint. In turn, poverty widely results in a high birth-rate, and consequent efforts to survive and cope with

---

* This updated version of 'Population, Economics, Development, and Environmental Security' (Tisdell, 1994) includes an extra case-study on biodiversity conservation in China.

93

population pressures can be expected to degrade the natural environment, result in its unsustainable use, and leave little or no margin for
the Human community involved to develop by its own efforts. The
local community is thus caught in a vicious cycle involving poverty,
population pressures, and environmental deterioration, so that escape
becomes more and more difficult with the passage of time. The desperate situation of the community may then become evident in terms
of increased frequency of famine, heightened mortality rates, and
outward migration.

However, increases in human population not only threaten the
natural environment and global sustainability in low-income
situations, but also pose a problem in higher-income situations. In
communities with higher *per caput* incomes, the incremental impact
of rising population on the natural environment tends to be greater
than in low-income situations, because the quantity of natural
resources transformed for human use, and the wastes and pollutants
generated by production and life-style maintenance, tend to be far
higher per individual. This really means that we must be aware of the
significance of the human population as an environmental influence,
no matter whether a community is poor or rich from an economic
viewpoint.

For at least another half-century, the world will have major
population problems to deal with. The German Advisory Council on
Global Change has summarized the situation as follows:

> 'During the 'nineties the global population is expanding annually by
> nearly 100 million people, most of whom (80 millions) will settle in
> urban areas. Even with a drastic decrease in the current average birth-
> rate, the present world's population of 5.52 [thousand millions] will at
> least double by the middle of the next century.
>
> 'The expected *population growth*, which will mainly occur in Asia,
> Africa, and Latin America, will aggravate both environmental and
> developmental problems. Conversely, however, the solution to these
> problems is a crucial prerequisite for the reduction of the population
> growth-rate. *Rapid urbanization* and *massive migration* are two further
> areas of concern, both of which will affect the industrialized countries
> because of increasing immigration pressure' (German Advisory Council
> on Global Change, 1993 p. 9).

The Council emphasizes that expected population increases will add
to environmental and development problems in the coming decades.
Therefore, there will clearly be a need for heightened concern and
policy effort to deal with such issues.

The importance which economists have placed on human
population-levels and their increase as impediments or facilitators of
economic development has varied over the centuries and continues
to vary also between individuals and factions. In addition, a variety

of views have been expressed about how or what socio-economic factors may influence family size and population growth, as well as about the likely environmental consequences of population growth. This Chapter provides an overview of population as an element in economic development models, of socio-economic influences on family size, of the dilemma raised by the 'population–environment–poverty trap' and responses to it — such as migration — and underlines the importance of wise natural-resource management.

POPULATION AS A VARIABLE IN ECONOMIC DEVELOPMENT MODELS

Population is an important variable in most economic development models, and population growth is seen by many (but not all) economists as a possible impediment to raising *per caput* incomes. Malthus (1798), for example, believed that population growth could be expected to fritter away the benefits of economic growth, whereas K. Marx and F. Engels (*cf.* Engels, 1973) were of the view that technological progress would more than compensate for any tendency of population growth to reduce *per caput* incomes. It is still the case today that those who are not especially concerned about the income-inhibiting consequences of population growth, point to technological progress and improvements in the quality of the work-force, *e.g. via* education, as important offsetting factors.

Malthus's population model has generally been interpreted to imply that there is always a strong tendency for the level of population to vary directly with the means of subsistence, so that in the long term *per caput* incomes hover around the prevalent levels of subsistence. However, the 'demographic transition' which has occurred in most Western countries since Malthus's time (and also in some non-Western ones) indicates the limitations of Malthus's theory (*cf.* Maitra, 1992). The natural rate of increase of population in most developed countries today is very small and in some cases negative, even though incomes per head are high. Indeed, high income levels often make for a reduction in the rate of population increase.

Such observations have led to the theory of the Malthusian 'population trap' or the low-level equilibrium of any subsistence-type 'population trap' (Todaro, 1981 Ch. 7). Basically, this theory holds that Malthus's model applies at low levels of *per caput* income but that if incomes can be raised enough, it is possible to escape from the Malthusian situation. At higher levels of income, population growth declines and, as a result of increased savings, higher capital investment, and related factors, forces resulting in rising incomes become stronger. Thus a situation of rising incomes and falling rates of population growth can be attained when once income per head rises to a sufficient level to escape the trap.

In simple theoretical renditions of this model, there is usually some threshold level of *per caput* income which is needed before self-perpetuating growth in *per caput* income and falling rates of population growth can be achieved. This of course is an over-simplified picture of the real world, but it does highlight the fact that a substantial improvement in the standard of living may be needed to escape a Malthusian population trap. In other words, the pres-cription for escaping from the trap may be an economic one, namely of economic development.

This point of view has been widely adopted by developmental policymakers. For example, the 'Brundtland Report' (World Com-mission on Environment and Development, 1987) accepted the view that economic improvement may be a prerequisite for many de-veloping countries to escape the Malthusian population trap and bring their rate of population growth under control. Such an escape is also seen to have environmental advantages.

This was pointed out in the 'Brundtland Report' and further underlined by the later World Bank report on *Development and the Environment* (World Bank, 1992). Taking a number of indices of environmental quality, the World Bank in this report indicates that, in general, environmental quality tends to be better in the more developed countries. This is so for water quality and urban-air quality, as examples. Furthermore, the higher life-expectancy in more-developed countries is an additional indicator of better environmental conditions in more-developed countries compared with less-developed ones.

However, the relationship between economic development and the state of the environment is complex. For example, at inter-mediate levels of *per caput* income, environmental quality is often worse than at low levels, because at the intermediate stage the worst spillovers from modernization may be observed and there may be a strong push 'to modernize' or industrialize at any (environmental) cost. If a country should become 'stuck' in the transition phase, this may be worse for the environment than if it had shown less eco-nomic growth.

In addition, value judgements are required in deciding what really is an improvement in environmental quality. As is to be expected, this in practice is judged mostly from a human per-spective. Often an improved environment from a human perspective means a substantially transformed natural environment.* Because of

---

*This reminds us of our — unpopular among conservationists — long-held con-tention that anthropogenic ecosystems and landscapes tend to be more attractive aesthetically, as well as far more productive economically, than natural ones. — Ed.

this and the increasing demand on resources resulting from economic growth and development, the natural environment tends to be
a casualty of 'development' unless positive measures are taken to
protect it. At higher-income levels it is economically easier than at
low ones to take such measures, but they do commonly require
positive governmental and often international action. *The developed
countries are, economically, in a favourable position to adopt
positive measures to protect their natural environment and to assist
less-developed countries in this regard.*

One of the main casualties of development is loss of biodiversity.
This is apparent from the World Bank (1992) as well as other
sources (Tisdell, 1993*b*). Both economic growth and increases in
human population are threats to biodiversity, and there again positive economic action is required to counter such threats.

Moreover, the thesis that economic growth and development lead
to an improved environment needs to be examined critically from
another point of view: to what extent have more-developed countries
or regions improved their environments locally by exporting their
environmental problems to less-developed ones? The developed
countries are, globally, the major contributors of 'greenhouse' gases,
and consequently export the risks involved to all nations. Furthermore, they sometimes locate their dirty industries offshore, or export
their wastes to less-developed countries (LDCs) where they may
cause serious environmental damage which would not be tolerated
in a more-developed country. Consequently, it must remain doubtful
whether the world, as a whole, really has the environmental capacity
to sustain its current population, let alone projected levels of population for the next 50 years or so, at high-income levels such as are
enjoyed for example in the USA and Switzerland on average.

What is being emphasized here is that cross-sectional data, such
as those used by the World Bank (1992), may give a misleading
picture of what is possible globally as far as the relationship between
development and the environment is concerned. Furthermore, one
has to be extremely careful in projecting past time-series data concerning the environment, as for instance is underlined by supporters
of the precautionary principle in relation to planning of development
and how to cope with environmental change.

The Malthusian population-trap theory has a significant consequence which has not been emphasized in the literature or in policymaking so far as I am aware. Premature and abortive attempts to
escape from such a trap may, in the longer term, do more harm than
good — because any growth achieved may prove basically non-
sustainable, and furthermore it is likely to reduce permanently the
natural resource-base, so making subsequent attempts at escape
more difficult.

For simplicity, suppose that a population is in a low-level-equilibrium situation and has a low *per caput* income level of y_1. Furthermore, suppose that if income per head can be pushed above $y_2 > y_1$, sustainable economic growth will occur in accordance with the Malthusian population-trap theory. Imagine that the community tries to escape from the trap (for example by drawing on its environmental resources or using some once-and-for-all foreign assistance) and its income per head rises to *almost* y_2 but cannot be raised further. Consequently, the community eventually returns to its low-level of income per head. In this equilibrium, it may only be able to support a smaller population than previously because of its irreversible natural-resource depletion during its attempt to escape from its low-level-equilibrium trap. Furthermore, because of environmental deterioration and natural-resource depletion due to the abortive attempts of the community to achieve sustainable growth, it is likely to be more difficult for such a community to escape subsequently from its low-level-equilibrium trap. Thus the theory suggests that 'premature' attempts to escape from such a trap should be avoided.

A further policy consequence is that, as far as foreign development-aid is concerned, it is best to concentrate it in areas where there is a high prospect of pushing communities above the Malthusian threshold level of income. This may not be achieved if aid is broadly dispersed or, alternatively, concentrated in regions where prospects of breaking out of the Malthusian trap are bleakest. Those regions which would find it easiest to escape from the trap should be assisted first, with less-promising regions being helped in their turn. This suggests that a serial approach to development assistance is needed if this theory is accepted. Although in the short-term this does not appear to be very humanitarian, in the longer term it may be the most effective policy holistically.

A very simple economic model of the relationship between population and income levels has not been mentioned. Basically, this model sees population as supplying labour to be used in the production process. The model concentrates on the production function to determine the optimal level of population, and most commonly the optimal level of population is taken to be that which maximizes income per head. As economies of scale in production are considered to be important, production as a function of population is assumed at first to exhibit increasing returns to rising population, followed eventually by decreasing returns. Consequently, this implies that the graph indicating income per head as a function of the level of population is of an inverted U-shape, such as is illustrated in Fig. 1. In the case illustrated, where income per head is shown by the curve OABC, the optimal level of population is P_1.

A population level of P_0 would be too low to maximize income per head, whereas one of P_3 would be too high.

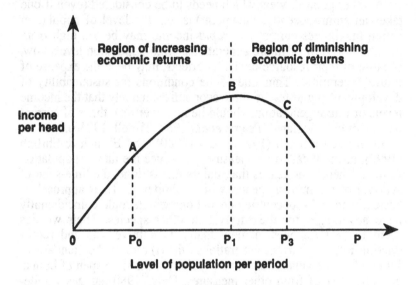

FIG. 1. *A commonly-assumed theoretical relationship between population level and income per head. It is based on the view that the productivity of a human population increases with growth in the size of the population at low levels of population, but eventually declines with increasing population because of economic constraints imposed by the limited availability of natural resources.*

While this rendition has some value, it presents a very static and partial view of population in relation to economic development, and takes population levels as being exogenously determined. Furthermore, it supposes that population is the only variable of significance for production. However, the relevant *per caput* income curve may shift, due to capital accumulation, technological progress, and/or greater education and knowledge. These shifts will usually be positive. But resource depletion and degradation may result in a negative or downward shift in the average income–population relationship.

Despite the limitations of the simple model depicted in Fig. 1, it highlights the question of what should be the criterion for determining an optimal level of human population. Should maximum *per caput* income of humans be the goal?

The answer depends in part on whether a purely anthropocentric view is taken. If expansion of human population after some point is at the expense of the survival of other species, and if it is believed that humans have some responsibility for preserving biodiversity, this would indicate that the optimal human population-level is likely to be somewhat lower than that which maximizes income *per caput* (Tisdell, 1990 Ch. 11, 1991 Ch. 11).

Another point of view which needs to be considered (even if one takes an anthropocentric viewpoint) is that the level of population which maximizes current *per caput* income may be too high from the point of view of future generations. High population levels now, because of their resource-depleting effects, may be at the expense of future generations. Thus one of the conditions for sustainability of development suggested by some economists, namely that the income levels of future generations should be no lower than those of current ones, would be violated (Pearce *et al.*, 1989; Tisdell, 1993*b*).

Georgescu-Roegen (1971), Daly (1980), and Ehrlich & Ehrlich (1990), have all called for measures to reduce the rate of population growth. Their basic fear is that, unless this is done, the time-span of survival of the human species will be shortened. Their approach is basically an anthropocentric one, but one which could coincidentally have advantages for the survival of other species. Their worries seem to be twin ones, namely that consumptive acts and repro-duction rates of current generations will (i) reduce the standard of living of future generations, and (ii) shorten the time-span of human existence. Apart from other measures, Daly (1980) suggests trade-able family-size permits as a way of regulating population increases. Without going into details, under such a system each person would be issued with a permit to have a specified number of children, *e.g.* one child; but the right would be tradeable. Thus a person wishing to have two children might purchase a permit from another person opting to have none.

Another problem needs to be mentioned, namely how is *per caput* income or the standard of living to be measured? The most common approach still is to use GDP (Gross Domestic Product) *per caput* as an indication of income per head and the standard of living. However GDP, as is well known, is grossly deficient as such an indicator: it does not include the value of non-marketed goods such as clean air, clear water, and access to harmonious natural sur-roundings. Furthermore, some environmental costs may be included as part of GDP because they lead to market transactions to avoid or reduce environmental damage. Such costs should be deducted from the welfare measure of the standard of living rather than added to it.

Many place considerable hope in new technology (*cf.* World Bank, 1992) as a way of overcoming prevailing constraints to economic development posed by rising population levels and con-comitant limitation of the natural environment. There is little doubt that such technology can be of great value in overcoming economic and environmental constraints. However, with the introduction of new technology, the scale of economic activity often expands.

Consequently, even if the new technology results in lower environmental damage per unit of activity, the expansion in the level of activity may result in the amount of *total* environmental damage actually rising. While many modern techniques are environmentally less damaging per unit of activity than earlier ones, the latter are still being used sometimes on an ever-expanding scale and with obvious environmental dangers.

As was pointed out above, a limitation of the simple static theory depicted in Fig. 1 is that it takes population as an exogenous variable, in contrast to the Malthusian theory and the Malthusian population-trap theories. In contrast to the Malthusian theory, Malthusian population-trap theories indicate that in some cases countries can escape *permanently* from subsistence equilibrium and population pressures.

In my view, Malthusian population-trap theories do deserve to be taken seriously. At the same time, it must be conceded that they are extremely crude. In particular, population growth depends on many factors other than levels of *per caput* income. This is clear from modern economic theories of family fertility and family size, which provide useful guidance on population policies and possible trends in population growth. Let us next consider the basic elements of these theories.

ECONOMIC THEORIES OF FAMILY FERTILITY AND FAMILY SIZE

Modern economic theory indicates that birth-rates and family sizes are significantly influenced by the economic circumstances of parents. One simple theory supposes that the sizes of families are largely determined by the desire of parents for children and the cost of having and raising them (Todaro, 1981 Ch. 7; Tietenberg, 1992 Ch. 5). This means that traditional economic theory can be adapted to analysing influences on family size. The demand curve of parents for children represents the value placed on extra children by parents in terms of their willingness to pay for extra children. In the case illustrated in Fig. 2, it is represented by the line marked DD.

The demand to have children will be high if local society puts considerable social value on large families, if children themselves can add to the productivity of the family as in some farming situations, or if they are essential for providing parents with income and security in their old age or if they should become handicapped. In many less-developed countries, especially in rural areas, the demand to have large families is still high because of one or more of such factors.

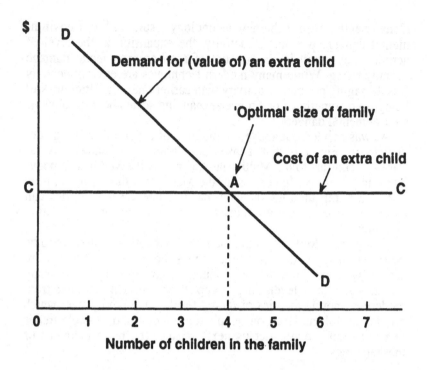

FIG. 2. *Family size determined by economic considerations, namely the value placed on an extra child and the cost of providing for it.*

The 'other side of the coin' is the cost of adding to one's family and maintaining children. In the case shown in Fig. 2, the additional cost to the family of an additional child is represented by the line marked CC. This may be relatively low in less-developed countries, because socially-required standards in provision of child-related services are usually lower in LDCs than in more developed countries. Furthermore, because of the extended-family system in most LDCs, and work being conducted in or near the home, as in many rural situations, children do not seriously interfere with the work-pattern of parents or require costly child-minding services as is the case in developed urban economies.

The size of the family, according to economic theory, is determined by the intersection of the curve showing the parents' demand for children and the additional costs involved in satisfying this demand. In Fig. 2, this occurs at the point of intersection marked 'A' and implies that the optimal size of the family from the point of view of these parents is four.

Any factor which adds to the cost of extra children will tend to reduce the size of the family. The disappearance of the extended-family system and urbanization are likely to do this — especially the former, which makes it more costly to parents to take care of children. Urbanization, where it involves work outside the home by men and women, also adds to the cost of child-care, and often children cannot be as productively employed in towns as in farming situations. Social pressures for education of children can also add to costs, and these pressures are likely to be stronger in an urban than a rural setting.

As societies develop and become more urbanized and market-oriented, the cost of having larger families rises, and the demand for, or value placed on, large families tends to fall. This trend is therefore a positive aspect of economic development from a population control viewpoint.

Furthermore, reductions in infant mortality and improvements in human health can be positive forces for reducing the birth-rate and average family size. When the probability of mortality is high among children, this can lead parents to have larger families or a greater number of childbirths to compensate for their possible loss of children after birth. Better health on the part of parents may, in addition, increase their productivity and income and reduce their need for assistance from their children, so favouring smaller-sized families.

Education, especially of women, has been identified in the literature as another positive influence in reducing family size. This may occur for a number of reasons: education may result in increased employment possibilities outside the home, improved social status (especially for females), higher incomes, and greater knowledge about methods and practice of contraception. All these factors make for families of smaller size.

Todaro (1981 p. 191) summarizes the economic theory of family size as follows:

> 'In deciding whether or not to have *additional* children, parents are assumed to weigh economic benefits against costs, where the principal benefits are, as we have seen, the expected income from child labor, usually on the farm, and their financial support for elderly parents. Balanced against these benefits are the two principal elements of "cost":
>
> 1. The "opportunity cost" of the mother's time (*i.e.* the income she could earn if she were not at home caring for her children).
>
> 2. The cost (both opportunity and actual) of educating children (*i.e.* the financial tradeoff between having fewer "high-quality", high-cost, educated children with high-income-earning potentials *versus* more "low-quality", low-cost, uneducated children with much lower earning prospects.'

Todaro (*ibid.*) goes on to elaborate on the theory and to draw policy conclusions from it. He says:

'Using the same "thought processes" as in the traditional theory of consumer behavior, the theory of family fertility as applied to LDCs concludes that when the price or cost of children rises as a result, say, of increased educational and employment opportunities for women, or a rise in school fees, or the establishment of minimum-age child labor laws, or the provision of publicly financed old-age social security schemes and so on, parents will demand *fewer* "additional" children, perhaps substituting quality for quantity or a mother's employment income for her childrearing activities. It follows that one way to induce families to desire *fewer* children is to raise the "price" of child rearing, by, say, providing greater educational opportunities and a wider range of higher-paying jobs for young women.'

Table I provides a summary of factors (mostly economic ones) which can be expected to lead to a reduction in family size. Most of these factors are positively correlated with economic growth, and can be correlated with economic development and a rising standard of living.

TABLE I

Factors Favourable to Reduced Family Size.

Demand-reducing Factors for High Birth-rates and Family Sizes	Factors Increasing Costs of Larger Families
– Social approval for small families. – Fewer opportunities for children to be productively employed. – Enhanced productivity of surviving children. – Availability of social security payments for old age or disability of parents. – Sustained health of parents, so less need for income support or other caring services.	– Rise in the social standards involved in the raising of children necessitating increased economic outlays, *e.g.* in relation to health-care and education. – Compulsory education of children where much of the cost of this education is borne by the parents. – Increased opportunity cost involved in child care, *e.g.* due to breakdown of the extended family, [and] employment of parents outside the home. – In some cases, economic penalties imposed by the State for large families (as in China), *e.g.* reduced work bonuses, increased family-support subsidies. – Increase in the cost of products used by children, *e.g.* food, clothing, and housing; medical expenses and costs for instance of schooling.

Other Influences Leading to Reduced Family Sizes

– Reduced cost of contraception, its greater reliability and acceptability.
– Increased knowledge of contraception techniques and family planning.
– Education appears to be positively connected with reduced family size.
– Improvement in the socio-economic status of women.
– Urbanization.

To regulate the size of their families, parents must know of suitable methods for this purpose and the methods must be within their economic reach. But this is not enough; they must also *want* to limit the size of their family, and exercise appropriate self-control and practised reticence. For economic or other reasons, parents may not desire a small family; consequently cheap contraception is not sufficient to ensure small-sized families, because it is only part of the decision process about family size. Self-discipline also plays a role.

The world is experiencing a rapid rate of urbanization, and this is especially apparent in LDCs. In fact, most of the increase in global population is expected to be accommodated in urban areas (German Advisory Council on Global Change, 1993). Such urbanization has been criticized as creating problems (Todaro, 1981) especially of an environmental nature. While this is true and will be a major challenge in the next few decades, urbanization also has a positive aspect in that it tends to result in smaller family sizes. This is because the costs of raising children in an urban environment are generally higher than in a rural environment, and usually the opportunities for productive employment of children are lower in towns — at least than in a countryside rich in farms.

THE NATURAL RESOURCE-BASE, ENVIRONMENT, AND POPULATION: SOME CASE-STUDIES

Particular cases enable us to consider ways in which rising population-densities can endanger the natural resource-base on which much economic production depends and likewise ultimately income-levels. Examples to be considered include shifting agriculture (which still occurs in parts of Asia, for example), farming in the Ukambani District of Kenya, and herding in the Sahel. They are indicative of a wide range of resource-problems which are emerging in rural areas owing to rising population-densities, and will now be dealt with in turn.

Shifting Agriculture

Shifting agriculture can take several different forms, but most of these forms involve a cultivation cycle where the land, after being cropped for a period of time, is allowed to return to a natural state of revegetation. The period of non-use of the land for agriculture allows for replacement of humus in the soil and a build-up of plant nutrients such as sources of nitrogen. Provided that the period allowed for natural revegetation of the cultivated land is long enough, the agricultural productivity of the land can be sustained (Ramakrishnan, 1987, 1992).

Nevertheless, as population densities rise, difficulties occur. The rotation cycle involving cultivation and fallow is altered, with the fallow-period becoming shorter in an attempt to feed more mouths by using the limited available land. Consequently, soil fertility tends to fall and with it the yields. Furthermore, the total proportion of land under cultivation at any one time rises. This usually results in an accelerated rate of soil erosion and, in wetter areas, greater water runoff. Thus flooding, siltation, and other unfavourable externalities, may occur with greater intensity or increased frequency downstream of the shifting cultivation, and with resultant economic and feeding-capacity loss. As yields decline on already cultivated land, economic pressures increase to bring even very marginal land (from an agricultural point of view) under cultivation, so adding to environ-mental degradation and unsustainable use of the natural resource-base, not to mention accelerated loss of biodiversity.

As a limit, shifting agriculture may give way to permanent agriculture. This may, however, not ensure sustainable yields or only do so (as practised) by the use of non-renewable resources, such as the application of nitrogenous fertilizers produced from natural gas. Furthermore, the local shifting cultivators may not have the know-ledge or culture which enables them to make a smooth transition to permanent agriculture. They may therefore be displaced by migrant agriculturalists who are accustomed to permanent agriculture. This has occurred in parts of the Indian subcontinent, *e.g.* displacement of hill tribes practising shifting agriculture in Bangladesh by Bengalis who are used only to lowland permanent agriculture. In the latter case, this has often added to environmental problems in hilly areas because the Bengalis have little experience with such cultivation and tend to transfer lowland farming practices to the hilly areas, thereby accelerating soil erosion (Tisdell *et al.*, in press).

In the case of tribal people following traditional methods of rural production, including shifting agriculture, the economic plight of many is not necessarily of their own making — that is, due to their own population increase. In many cases, they are suffering en-croachment on, and loss of, their traditional lands as a result of the economic expansion and population increase or explosion of other groups. This may take the form of actual appropriation of their traditional lands for farming by a dominant group, or the use of those lands for the provision of infrastructure such as dams and reservoirs for servicing the economic activities of the dominant group. For example, the building of the Kapatur Dam in the Chittagong hill tracts of Bangladesh meant the loss of a considerable amount of land by local hill-tribes. Thus the population density on the remaining land area of hill tribes in the Chittagong hill tracts increased (Tisdell *et al.*, in press). Furthermore, these losses are compounded as Bengali farmers increasingly settle in the foothills of this region (Alauddin *et al.*, 1995). This pattern is not, however, by any means unique to Bangladesh.

The acquisition (appropriation) of tribal lands by those practising more intensive forms of agriculture and rural production raises few moral qualms for those doing the appropriation. They tend to look upon such lands as wastelands or neglected lands, and consider morality to be on their side in raising the economic productivity of these lands by appropriating them and using them more intensively. However, this enhanced productivity may not be sustainable. Furthermore, those acquiring such land and using it more intensely than hitherto are likely to fail to take account of any unfavourable externalities that arise from intensification of economic activity on the land. Thus not only are tribal people marginalized, but those replacing them may also be headed for poverty due to the adoption of unsustainable land practices.

It might be noted that the appropriation of land used by tribal people and by those engaged in extensive rural land-uses, or by others who are intent on its intensive use, is not only made easier by the above moral view but also by the fact that the extensive users of the land often have no formal legal title to it. In addition, they are often not well educated and competent at making legal claims, which adds to the chances of their dispossession. Furthermore, the dominant economic group usually controls the legislature, and so expropriation may in fact proceed legally, *e.g.* all 'vacant' land may be made state property and redistributed.

It thus seems on the whole that those societies which engage in the most intensive forms of economic activity become wealthier and more powerful compared with those which are reliant on more extensive forms of economic activity, though the latter are often in greater harmony with Nature. Consequently, there has been a tendency for societies that are engaged in intensive economic activity to supplant those which are based on extensive economic activity. While 'frontiers' in which land was little-used remained, these changes could be accommodated to some extent by the emigration of those involved in the more extensive forms of economic activity. However, most such 'frontiers' have long since disappeared, so the brunt of population increase and intensification of economic activity by dominant groups now weighs very heavily on those practising extensive forms of economic activity. Of this an example is given later in this Chapter regarding nomadic herding in the Sahel.

Unsustainable Farming in the Ukambani Region of Kenya

The Akamba people live in the Machakos and Kitui districts (Ukambani) in central-southern Kenya. The area consists of low hills bordering onto plains and is very 'marginal' for agriculture. The Akamba were originally herders and shifting cultivators in more productive areas of Kenya but were displaced by the Maasai by force in competition for land and retreated to the Ukambani region. As a result of loss of land to the Maasai, and their own rising population, the Akamba have been forced to alter their land-use from one of shifting agriculture and herding to one of settled agriculture. However, yields from this agriculture are unsustainable in Ukambani.

The situation has been explained by Lee (1993 pp. 10–1) in the following terms:

'Towards the end of the 19th century, skirmishes between the Akamba and the Maasai subsided. As their population increased, the Akamba expanded onto the surrounding plains. Initially, they used the plains for cattle herding, but the need to grow grain brought increasing cultivation around their homesteads.

'To begin with, the farmers practised shifting cultivation. As each culti-vated site became less productive, due to soil depletion, they moved to another site. But as the population pressure increased during this century, opportunities for shifting cultivation onto new desirable land diminished. The Akamba thus became more intensive cultivators on fixed locations, eventually with fixed land-titles. Today the rate of population increase in Ukambani is one of the highest in the world, but the productivity of the soil continues to fall.

'The social and economic problems of the area have been of concern for more than 70 years. There have been several cycles of rapid popu-lation growth, soil depletion, and erosion, and attempts to solve the problem. These have included emigration to new settlement areas, terracing to protect the land from erosion, and introduction of an early-maturing maize variety, called 'Katumani Composite', that produces more than the traditional maize varieties in poor seasons. Nevertheless, soil fertility has continued to decline.'

The Akamba farmers have thus been caught in a poverty trap in which population increase and economic circumstances have com-bined to ensnare them more and more firmly. Many different cycles are involved. The area under cropping is increasing as available land is subdivided and parcelled out to the rising population. This means that the area for livestock is shrinking and stocking densities are growing, leading to overgrazing and the need to supplement the food supply of cattle. In fact a soil degradation cycle is involved which is increasingly difficult to break out of.

The soil degradation cycle involves the following factors:

(1) Soil organic matter is declining because crop residues (mostly of Maize, *Zea mays*) must be fed to the animals and manure from the 'bomas' (livestock corrals) is not returned to the fields, mainly because of the effort involved.

(2) Because of the reduction of organic matter in the soil, useful biological processes in it are impeded and there is low infiltration of rain-water, with resultant accelerated runoff.

(3) Consequently there is a loss of soil and nutrients and crop yields fall. Hence, the incomes of the Akamba farmers are reduced and the incidence of poverty rises.

The course of this downward economic spiral is indicated in Fig. 3.

The problems faced by the farmers of Ukambani have been investigated by the Kenyan Agricultural Research Institute and Australian researchers in the CSIRO (Commonwealth Scientific and Industrial Research Organization). Their recommendation for breaking the poverty cycle is, for the Akamba:

(1) to apply commercial fertilizer in small quantities;

(2) preferably to spread manure from the bomas on the fields and include legumes* in their crop rotation; and,

(3) from the increased yield of crops that have been attended to in this way, to reserve a part of the coarser crop residues, such as corn-stalks, for return to the soil so as to provide more humus.

The suggestion is a positive one. Nevertheless, it is not certain to result in escape from the Malthusian population-trap. Its success depends on farmers supplementing their limited use of fertilizer by other sustainable practices, *e.g.* return of a portion of the increased crop residues to the soil, spreading of manure, and crop rotation using legumes.* Pressure to feed increased crop residues to livestock may be strong, and rising population levels may add to this pressure. The prospects of escape from the 'trap' may therefore be limited. Furthermore, the fertilizer will require imports, adding to Kenya's balance of payments difficulties. Further yields may become fertilizer-dependent, and in the longer term be sustained only by increasing applications of chemical fertilizer (*cf.* Tisdell, 1991 Ch. 9; 1993*a* Ch. 11). Nevertheless, if the full set of recommendations were adopted and population levels could be stabilized, the proposals would be of great assistance.

Unfortunately, there are not promising prospects for stabilizing the population in Ukambani, where females have little opportunity for off-farm employment. Although males often go to Nairobi and other urban centres for work, their womenfolk are left in the villages

---

* Members of the pea and bean family (Leguminosae), commonly fixing atmospheric nitrogen and hence of nutritional value to crops, increasing their yield. — Ed.

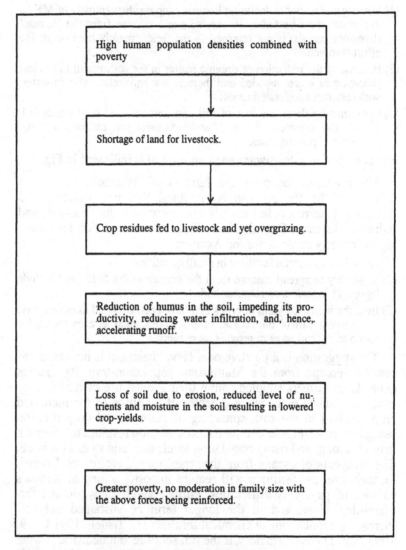

FIG. 3. *A cycle induced by poverty and population-pressure involving soil degradation. An example of a Malthusian population-trap.*

and primarily take care of the farming, with the males returning periodically. This form of partial urbanization does little to put a brake on population growth.

The Nomadic Herders of the Sahel

Another group of extensive land-users who have come under growing resource and economic pressures in recent decades are nomadic herders of livestock. This has occurred in many parts of the globe and is particularly evident in the Sahel region of northern Africa.

As population levels in the Sahel have increased, this has led to an expansion of agriculture, often supported by foreign aid, at the expense of nomadic herding. Increasingly 'fixed' agriculturalists have appropriated the traditional grazing-lands of nomads — the land which the latter had long used for grazing and access to vital drinking-water for their livestock. This is well documented by cases described in Bennett (1991).

Several consequences have followed, which include:

(1) Greater crowding of livestock on the reduced total area of land available for nomadic grazing, with accelerated environmental deterioration thereof.
(2) Greater degree of poverty and frequency of famine.
(3) Ecologically induced migration of nomads to urban areas where they often become fringe-dwellers and have great difficulty in adjusting to the urban situation.
(4) Increased violence and armed conflicts between land-users — that is, between nomadic pastoralists and settled agriculturalists, and between competing nomadic groups or tribes.
(5) Ecologically unsustainable use of the land by agriculturalists or by those engaged in pastoralism in a fixed location.

This case of nomadic herders, as well as those described above, illustrate the dangers inherent in intensification of rural land-use. One must carefully consider whether increased yields from such intensification are ecologically sustainable. It is not that intensification of land-use should never be contemplated as a strategy for economic development, but that it is necessary to consider very carefully whether a particular case is sustainable and will show a reasonable degree of internal ecological balance — bearing in mind that systems which have been heavily reliant on outside energy resources may not be sustainable in the long run.

Biodiversity Conservation in China: a Brief Case-study

China provides yet another case of conflict between environmental conservation, economic growth, and population increase. Since 1978, when China commenced its economic reforms designed to increase the use of market forces in its economy, its economic growth has been faster than that of most countries (Tisdell, 1993*d*). However, this growth, coupled with continuing population increases, has not been without environmental costs. Primarily as a result of population expansion and economic growth, China has experienced a substantial reduction in the extent of its natural forests, woodlands, wetlands, and grasslands, *e.g.* due to the encroachments of agri-

culture onto naturally vegetated areas, the intensification of agriculture, and the need for fuel and timber supplies. Although China
has a one-child policy, this policy does not apply to minorities and is
difficult to enforce in rural areas. So population increase in many
rural areas continues to be a matter of concern.

Apart from that resulting from the loss of natural vegetation
cover, the introduction of aggressive exotic species to China and the
adoption of high-yielding varieties of agricultural crops have all led
to reduced biodiversity in China. Whereas 10–15% of species of
plants are considered to be threatened or endangered globally, in
China 15–20% of its species are believed to be threatened or
endangered nationally, which is proportionately up to twice the
world's average. Consequently, a high priority needs to be given to
biodiversity conservation in China. This is recognized in *China's
Agenda 21 — White Paper on China's Population, Environment and
Development in the 21st Century* (State Council, 1994) and in
China: Biodiversity Conservation Action Plan (National Environmental Protection Agency and others, 1994). Both of these documents represent a search for strategies to ensure environmentally
sustainable development and the maintenance of biodiversity, but
not easy solutions appear to be available.

A major problem is how to provide adequate economic benefits
to local people without compromising the maintenance of biodiversity. The problem can be illustrated briefly by Xishuangbanna
Dai Autonomous Prefecture located in southern Yunnan. This
Prefecture borders on Laos and Burma. According to Myers (1988,
1990) and Mittermeier & Werner (1990), this is a region of
'biological megadiversity' and is thus deserving of a high priority
for conservation purposes. It has a rich variety of plants and contains
the only remaining populations of the Asiatic Elephant (*Elephas
maximus*) in China.

However, its natural environment is under continuing threat and
subject to considerable modification. The rate of human population
growth of this Prefecture is rapid, due to natural increase as
minorities predominate outside of its towns. Currently it is reported
that there is little inward migration to this Prefecture, though prior to
1978 considerable numbers of Han Chinese were relocated in this
area. Rising rural populations have resulted in intensification of
land-use outside of protected areas. Permanent agriculture has en-
croached onto ever-more marginal land and the cycles of shifting
agriculture, practised by some minorities, have become shorter.
These factors, combined with increased demands for fuel and timber, have — apart from increasing rates of soil erosion — severely
reduced forested and naturally-vegetated areas in the Prefecture, as
is clear from comparative satellite photographs. Furthermore, human

pressures on protected areas within this Prefecture have mounted. Illegal hunting of animals, and removal of plants, timber, and so on, are continual problems that are exacerbated by the rising population and denudation of land areas outside of reserves and protected areas.

Strategies that are being employed, or are being considered by the Chinese government, for reducing human pressure on protected areas in this Prefecture include:

(1) The encouragement of ecotourism employing the 'user-pays' principle and involving local communities. For instance, a joint local venture has been established with a local village near Menglun to open a portion of Xishuangbanna State Nature Reserve to ecotourism.

(2) Granting of limited rights to local communities to use protected areas for extractive purposes or for restricted cultivation. For example, rights are given to some local communities within Xishuangbanna State Nature Reserve to cultivate a local 'ginger plant' that is used for medicinal purposes. This plant grows only on the forest floor, and forest use which has a minimally adverse impact on conservation can thus be legitimized.

(3) Government financial support for economic development projects in villages bordering on or located near protected areas. It is hoped that such projects, by raising the incomes of villagers, will make them less inclined to use protected areas illegally. Where this project is subsidized by the government, the subsidy can provide a lever for compliance with the law; for example, the subsidy may be withdrawn in the event of any illegal use of protected areas.

(4) Some other methods of law-enforcement include small payments to village leaders to act as 'forest guards'.

Whether or not such measures will work remains to be seen. Although the prospects cannot be discussed here, it should be noted that increasing population mobility in China could reduce human populations located in or near protected areas and so reduce pressures from this immediate source. Unfortunately, however, pressures on protected areas do not only come from local populations. For example, growing urban populations, industrialization, and 'modernization', add to demands for greater energy supplies, and there is rising demand in China for electricity, including in the urban areas of Xishuangbanna. In Xishuangbanna, electricity supplies could be expanded by constructing a dam on the Lancang River (known downstream as the Mekong River). Such a dam could also supply irrigation water for agricultural use in the dry season. But while the possibility has been seriously considered by China, no decision has yet been taken as to whether to build this proposed dam.

A major disadvantage of such a dam (given the suggested site for it) would be that it would flood a large part of Xishuangbanna State Nature Reserve and further threaten the maintenance of biodiversity in this Prefecture. In fact, the building of infrastructure throughout

China (and for that matter throughout most of Asia) as a part of its drive for 'modernization', and its efforts to raise living standards, constitutes a major threat to the preservation of biodiversity. The road to ecologically sustainable development in Asia is not likely to be easy, particularly if the maintenance of biodiversity is considered to be a desirable part of it. In Asia, urbanization is accelerating and its consequences for the environmental future need to be increasingly considered, taking into account the ecological and economic interdependence of urban and rural areas (Tisdell, 1995). Rural environments are all-too-often altered or sacrificed to satisfy the demands of urban growth, and in such cases we must look beyond rural communities to establish trends.

CONCLUDING COMMENTS

Since the time of Malthus's epoch-making *Essay* (1798), there has been considerable progress in forming an effective economic theory of population. The modern economic theory of optimal family size has proven to be useful for the purposes of both prediction and policymaking. There is an increasing need to link this theory with the prospects for environmentally sustainable development and for escape from poverty cycles that are characteristic of many LDCs, involving, as most of them do, the economically unsustainable use of natural resources and concomitant environmental degradation.

We have been able to explore the difficulties of escaping from a low-level equilibrium 'trap' involving population, poverty, and natural resources, in a general context as well as for particular cases. The particular cases, however, are not unique but are chosen to illustrate the difficulties of combining economic growth and population growth with a secure environmental future. They underline the importance of *fully exploring ecological consequences* of changes in land-use before actually embarking on such changes, so as to ensure wise resource management, sustainable land-use, and ultimately economically sustainable incomes and conditions with good prospects for population control.

REFERENCES

ALAUDDIN, M., MUJERI, M.K. & TISDELL, C.A. (1995). Technology–employment–environment linkages and the rural poor of Bangladesh: insights from farm-level data. Pp. 221–5 in I. AHMED & J.A. DOELEMAN (Eds) *Beyond Rio: The Environmental Crisis, Technological Dilemma and Sustainable Livelihoods in the Third World*. Macmillan, London, UK: xx + 390 pp., illustr.

BENNETT, O. (Ed.) (1991). *Greenwar: Environment and Conflict*. Panos Institute, London, England, UK: iv + 156 pp., illustr.

DALY, H.E. (1980). *Economics, Ecology and Ethics: Towards a Steady-state Economy.* Freeman, San Francisco, California, USA: xii + 372 pp., illustr.

EHRLICH, P.R. & EHRLICH, A.H. (1990). *The Population Explosion.* Simon & Schuster, New York, NY, USA: 320 pp.

ENGELS, F. (1973). Outlines of a Critique of Political Economy. In K. MARX *Economic and Philosophic Manuscripts of 1844,* Foreign Languages Publishing House, Moscow, Russia, also cited as 1973, Lawrence & Wishart, London, England, UK: 255 pp., but not available for checking.

GEORGESCU-ROEGEN, N. (1971). *The Entropy Law and the Economic Process.* Harvard University Press, Cambridge, Massachusetts, USA: xv + 457 pp., illustr.

GERMAN ADVISORY COUNCIL ON GLOBAL CHANGE (1993). The world in transition: basic structure of the global Man–Environment relationship. *Global Change Prisma,* 4(2), pp. 7–10.

LEE, B. (1993). *Escaping from Hunger.* Australian Centre for International Agricultural Research, Canberra, Australia: 52 pp., illustr.

MAITRA, P. (1992). The demographic effects of technological change and capitalist transformation in a re-interpretation of the Demographic Transition Theory. *Artha Vijana,* 34(2), pp. 125–54.

MALTHUS, T.R. (1798). *An Essay on the Principle of Population as it Affects the Future Improvements of Society.* [Reprint 1909. Macmillan, London, England, UK: xix + 134 pp.]

MITTERMEIER, R.A. & WERNER, T.B. (1990). World of plants and animals in 'megadiversity' countries. *Tropicus,* 4(1), pp. 4–5.

MYERS, N. (1988). Threatened biota: hotspots in tropical forests. *The Environmentalist,* 8(8), pp. 1–20.

MYERS, N. (1990). The biodiversity challenges: expanded hot-spot analysis. *The Environmentalist,* 10(4), pp. 243–56.

NATIONAL ENVIRONMENTAL PROTECTION AGENCY AND OTHERS (1994). *China: Biodiversity Conservation Action Plan.* National Environmental Protection Agency, Beijing, China: v + 106 pp.

PEARCE, D., MARKANDYA, A. & BARBIER, E.G. (1989). *Blueprint for a Green Economy.* Earthscan Publications, London, England, UK: xvi + 192 pp., illustr.

RAMAKRISHNAN, P.S. (1987). Shifting agriculture and rainforest ecosystem management. *Biology International,* 15, p. 17.

RAMAKRISHNAN, P.S. (1992). *Shifting Agriculture and Sustainable Development: An Interdisciplinary Study from North-eastern India.* UNESCO, France–USA–England: xvii + 424 pp., illustr., tables.

STATE COUNCIL (1994). *China's Agenda 21 — White Paper on China's Population, Environment and Development in the 21st Century.* China Environmental Science Press, Beijing, China: iv + 246 pp.

TIETENBERG, T. (1992). *Environmental and Natural Resource Economics,* 3rd edn. Harper Collins, New York, NY, USA: xxvi + 678 pp., illustr.

TISDELL, C.A. (1990). *Natural Resources, Growth and Development.* Praeger, New York, NY, USA: xvi + 187 pp., illustr.

TISDELL, C.A. (1991). *Economics of Environmental Conservation.* Elsevier Science Publishers, Amsterdam, Netherlands: xxii + 234 pp., illustr.

TISDELL, C.A. (1993a). *Environmental Economics.* Edward Elgar, Aldershot, England, UK: xii + 260 pp., illustr.

TISDELL, C.A. (1993b). *Combining Biological Conservation, Sustainability and Economic Growth: Can We Overcome Potential Conflict?* Discussion Paper No. 130, Department of Economics, University of Queensland, Brisbane 4072, Australia: 11 pp. (An invited keynote paper for the 20th Waigani Seminar on Environmental Development, 23–27 August 1993, Port Moresby, Papua New Guinea.)

TISDELL, C.A. (1993c). Ecological economics and the environmental future. Pp. 363–76 and following Commentary in *Surviving With The Biosphere: Proceedings of the Fourth International Conference on Environmental Future (4th ICEF), held in Budapest, Hungary, during 22–27 April 1990* (Eds NICHOLAS POLUNIN & Sir JOHN BURNETT). Edinburgh University Press, Edinburgh, Scotland, UK: xxii + 572 pp., illustr.

TISDELL, C.A. (1993d). *Economic Development in the Context of China.* Macmillan, London, UK: xiv + 220 pp., illustr.

TISDELL, C.A. (1994). Population, economics, development, and environmental security. Pp. 63–84 in *Population and Global Security* (Eds NICHOLAS POLUNIN & MOHAMMAD NAZIM), *Limited Geneva Edition for* United Nations Population Fund (UNFPA) *et al.* The Foundation for Environmental Conservation, Geneva, Switzerland, & Energy and Environment Society of Pakistan, Lahore, Pakistan: xi + 285 pp., illustr. and tables.

TISDELL, C.A. (1995). Asian development and environmental dilemmas. *Contemporary Economic Policy,* **13,** pp. 38–49.

TISDELL, C.A., ALAUDDIN, M. & MUJERI, M. (in press). Water resource availability, management and environmental spillovers in Bangladesh. In *Perspectives in Resource Management in Developing Countries* (Ed. A. Thakur). Concept Publishing Company, New Delhi, India.

TODARO, M.P. (1981). *Economic Development in the Third World,* 2nd edn. Longman, New York, NY, USA: xxxiv + 588 pp., illustr.

WORLD BANK (1992). *World Development Report 1992: Development and the Environment.* Oxford University Press, New York, NY, USA: xii + 308 pp., illustr.

WORLD COMMISSION ON ENVIRONMENT AND DEVELOPMENT (1987). *Our Common Future.* Oxford University Press, Oxford, England, UK: xv + 400 pp., tables.

6

Imperatives for Environmental Sustainability: Decrease Overconsumption and Stabilize Population*

by

Robert J.A. Goodland, PhD (McGill),
*Environment Department,
The World Bank,
1818 H Street,
Washington,
DC 20433,
USA,*

&

Herman E. Daly, PhD (Vanderbilt),
*School of Public Affairs,
University of Maryland,
College Park,
Maryland 20742-1821,
USA.*

Introduction

If we examine the state of the world as it approaches our twenty-first century AD, we find humanity hurtling away from sustainability at an unprecedented rate. Unbridled consumerism in developed nations, marked by increased throughput of environmental resources, compete with escalating population growth in the South as principal threats to any serious notion of real sustainability (Daily & Ehrlich, 1992; Hardin, 1993). This Chapter, updated and finalized after the United Nations' Conference on Population and Development was held in Cairo in September 1994, examines

* The opinions expressed in this Chapter are the personal views of the Authors alone, and should in no way be construed as reflecting the official position of The World Bank Group.

whether humanity can regain a standard of living which does not cause it to exceed the carrying capacity of the planet. It is our belief that the transition to environmental sustainability can and *will* occur. However, the ability and real desire of nations to plan for an orderly transition to environmental sustainability, rather than allowing bio-physical limits to dictate the timing and course of such a necessary transition, remains in question.

What is the nature of environmental sustainability? This Chapter refers to *Environmental* Sustainability, not to the related but sub-servient concepts of Social Sustainability and Economic Sustain-ability, contrasted in Table I. But how can environmental sustain-ability be attained and what are its characteristics? These are among the issues outlined in this Chapter, which seeks to develop a frame-work for addressing the transition towards environmental sustain-ability as we conceive it and believe our world of Humankind and Nature needs it.

One measure of the impact of human activities upon a local resource-base, synonymous with environmental sustainability, is given by Ehrlich & Holdren's equation I = PAT. They noted (Ehrlich & Holdren, 1974) that the impact of any population or nation upon environmental sources and sinks* is a product of its population (P), its level of affluence (A), and the damage done by the particular technologies (T) which support that affluence:

$$I = P \times A \times T$$

In this instance, affluence is measured by *per caput* consumption of resources. The impact due to increased levels of affluence consists of the material flows needed to maintain each form of capital. Technology refers to technological efficiency defined in terms of the number of units of output or consumption produced per unit of environmental cost. Population, affluence, and technology, are not unrelated. For example, technology varies as a non-linear function of population, affluence, and rates of change, in each of these factors; more powerful technologies are needed to provide more people with higher levels of affluence. Throughput of matter and energy must stay within the carrying capacity of the local environs (*i.e.* throughput < carrying capacity).

Estimating the product of *per caput* consumption of resources and technological efficiency within the I = PAT framework is

---

* Early studies of environmental limits to human activities emphasized the limits to environmental resources (*i.e.* petroleum, copper, etc.) (Meadows *et al.*, 1972). Experience has shown, however, that the 'sink constraints' (*i.e.* waste assimilation such as air and water pollution, 'greenhouse' gases, and stratospheric ozone depletion, etc.) are apt to be more stringent (*ibid.*).

TABLE I

Comparison of Social, Economic, and Environmental, Sustainability.

Social Sustainability ('Soc. Sust.')	Economic Sustainability (EcS)	Environmental Sustainability (ES)
Socio-cultural stability is the social scaffolding provided through networks of peoples' organizations that empower self-control and self-policing in peoples' management of natural resources. Resources should be used in ways which increase equity and social justice, while reducing social disruptions. Human rights, education, employment, women's empowerment, and openness in negotiation and participation, seem to be integral to Soc. Sust. It will emphasize qualitative improvement of social organization patterns over quantitative growth of physical assets, and cradle-to-grave pricing to cover full costs — especially social costs.	Economic capital should be stable. The widely accepted definition of economic sustainability — 'maintenance of capital', or keeping capital intact — has been used by accountants since the Middle Ages to enable merchant traders to know how much of their sales receipts they and their families could consume without reducing their ability to continue trading. Thus Hicks' (1946) definition of income — *the amount one can consume during a period and still be as well off at the end of the period'* — can define EcS, as it devolves on consuming interest rather than capital.	Although environmental sustainability is needed by humans and originated because of social concerns, ES seeks to improve human welfare by protecting the sources of raw materials which are used for human needs and ensuring that the sinks for human wastes are not exceeded, in order to prevent harm to humans.
Thus social science should have much to contribute to ES. However, Soc. Sust. will be achieved only by systematic community participation from a strong civil society. Social cohesion, cultural identity, diversity, solidarity, sense of community, tolerance, humility, compassion, patience, forbearance, fellowship, fraternity, established institutions, love, pluralism, commonly accepted standards of honesty, laws, discipline, etc., constitute the part of social capital that is least subject to rigorous measurement, but is widely most important for Soc. Sust.	We now need to extrapolate the definition of Hicksian income from sole focus on human-made capital and its surrogate (money) to embrace the other three forms of capital (natural, social, and human). Economics has rarely been concerned with natural capital (*e.g.* intact forests, healthy air) because until relatively recently it had not been scarce. This new pattern of scarcity, namely of scarce natural capital, arose because the scale of the human economic subsystem has latterly grown large relative to its supporting ecosystems and wider ecocomplexes. To the traditional economic criteria of allocation and efficiency, must now be added a third, namely that of scale (Daly, 1992). The scale criterion would constrain throughput growth — the flow of material and energy (natural capital) from environmental sources to sinks, *via* the human economic subsystem.	Humanity must learn to live within the limitations of the biophysical environment. ES required that natural capital must be maintained, both as a provider of inputs ('sources') and as a 'sink' for wastes (Daly, 1972; Daly & Cobb, 1974; Serageldin, Daly & Goodland, 1995). This means holding the scale of the human economic subsystem to within the biophysical limits of the overall superecocomplex (Planet Earth) on which it depends. ES needs both sustainable production and sustainable consumption.
This 'moral capital' requires maintenance and replenishment by shared values and equal rights as well as by community, religious, and cultural, interactions. Without this care it will depreciate just as surely as will physical capital. Human capital — investments in education, health, and nutrition, of individuals — is now accepted as part of economic development, but the creation of social capital, as needed for Soc. Sust., is not yet adequately recognized.	Economists prefer to value things in terms of money, and are having major problems in valuing *natural* capital — intangible, intergenerational, and especially common-access resources, such as air, etc. In addition, environmental costs used to be 'externalized', but are now starting to be internalized through sound environmental policies and valuation techniques. Because people and irreversibles are at stake, economics needs to use anticipation and the precautionary principle routinely, and should err on the side of caution in the face of uncertainty and risk.	On the 'sink' side, this translates into holding waste emissions within the assimilative capacity of the environment without in any way impairing it. On the 'source' side, harvested rates of renewables must be kept within assured regeneration rates. Non-renewables cannot be made fully sustainable, but quasi-ES can be approached for non-renewables by holding their depletion rates equal to the rate at which renewable substitutes can be created (*see* Serafy, 1991, 1993). Ultimately, *there can be no* Soc. Sust. without ES, as ES supplies the conditions for Soc. Sust. to be approached and hopefully attained.

difficult, and thus *per caput* energy use is frequently employed as an imperfect surrogate. Using that crude measure, and dividing rich and poor nations at a *per caput* GNP of $4,000, each human inhabitant of the developed world does roughly 7.5 times as much damage to the Earth's life-support systems as does an inhabitant of the third or less-developed world (Goodland *et al.*, 1993). At the extremes, the impact of a typical person in a desperately poor country is roughly one-thirtieth that of an average citizen of the United States (*cf.* Chapter 2 for documentation). Sustainability would be approached far sooner if the USA were to achieve an average family size of 1.5, than would a similar success in Bangladesh (Ehrlich, 1994). OECD's 5% of global population growth, and OECD consumption levels, burden the environment more than the 95% of other population growth in low-consumer countries. The United States now has the fastest-growing population in OECD, and immigration accounts for a large fraction of that rate.

The conclusions to be drawn from the I = PAT equation are clear. There are only three means of reducing the impact of human activities upon an already-stressed environment. These are: (1) limiting population growth, particularly in developing nations; (2) limiting affluence, particularly in developed nations; or (3) improving technology, thereby reducing throughput intensity (Goodland *et al.*, 1993).

LIMITING POPULATION

In late-1993, the world's human population stood at 5.6 thousand millions. More than one thousand millions of these people — roughly one-fourth of the total population of developing nations — lived below the poverty line, and 100 millions of these individuals lived in abject poverty, lacking even the most basic necessities of food, clothing, and housing. In most countries in Africa, and several in Latin America, *per caput* food consumption is still falling and living standards are deteriorating (*cf.* Chapter 2). In the face of these conditions, the world population is now projected to increase by 84 millions annually — approximately the entire population of Mexico. Of these, some 80 millions are citizens of so-called 'developing nations', and most of these will live in ecologically vulnerable areas.

While the global population growth-rate is 1.7%[§] annually, the combined population of less-developed nations (excluding China) is growing at a rate of 2.4% annually and is projected to double in approximately 30 years' time if no changes in fertility or mortality-rates occur. The annual rate of population growth in more-developed

[§] Early in 1997 reported to be reduced to 'just under 1.5'% — *see* Chapter 3. — Ed.

nations is 0.5% (*cf.* Chapter 2). Because the 'impact' of one Northerner* exceeds that of one Southerner[†] by an order of magnitude of some ten times, Northern populations must stabilize quickly. Even 1% population growth-rate doubles the population in 70 years; and the 'population' of Northern artefacts (*e.g.* private automobiles) must stabilize or decline even faster, given that the impact of one car exceeds the impact of one human — even a Northerner — also by an order of magnitude or more. But 90% of future population growth is predicted to occur in low- and middle-income nations (*cf.* Chapter 2), so population control also is needed there. Indeed, rapid population growth within subsistence economies compounds environmental degradation (*i.e.* soil erosion, depletion of natural capital, deforestation and its indirect effects on watersheds, irrigation, etc.), thereby increasing the impact of human activities on local environs (Daily & Ehrlich, 1992).

Slowing the momentum of population growth in low- and middle-income nations will not be an easy task. Whereas the mean proportion of the population under 16 years of age in high-income nations is 21%, in low- and middle-income nations (excluding China) this figure approaches 40%. In many nations in Africa, nearly 50% of the population are under the age of sixteen. Age-structures so heavily skewed towards the young, who have yet to live their reproductive years, generate tremendous demographic momentum, particularly as modern medicine contributes to lowering death-rates (especially infant mortality) in developing nations.

In the transition toward environmental sustainability, *population stabilization is the only acceptable goal*. Indeed, if we cannot provide for 20% of the human population at today's relatively low population, how will we be capable of maintaining living standards when (if) the world population surpasses ten thousand millions?[§] Thus, within low- and middle-income nations, population stabilization will require families to be limited to 2.5 to 3.0 children, whereas in industrialized nations, with lower infant mortality rates, replacement-level fertility is approximately 2.1 children per family. The means to approach population stability are outlined in Table II (p. 129).

---

* Meaning, approximately, an inhabitant of a developed or industrialized, 'First World' country, regardless of its geographical position. — Ed.

[†] Meaning, approximately, an inhabitant of a less-developed or developing, 'Third World' country, regardless of its geographical position. — Ed.

[§] Technological optimists hope that technological improvements will appear in time to allow us to cope with such population growth. This is dangerously misleading (*see* our section below on Improving Technology) if recent history is any guide.

Leading priorities for population stabilization include:

1. Meet the unmet demand for family planning services. As many as 300 million married women* of reproductive age would like to prevent pregnancies but have no access to reliable birth-control. Few if any efforts to improve the human condition will pay a higher return on investment than democratizing access to contraception if the global good is seriously considered (Goodland *et al.*, 1993).

2. Support information campaigns directed towards women *and* men, particularly addressing the links between family size and economic impoverishment, and between human population and the state of the world‡.

3. Increase educational opportunities for girls and young women, as well as employment opportunities for women. As incomes rise and education and employment opportunities for women expand, fertility levels (in the sense of the number of children per woman) decline markedly (*see* Chapter 2 for documentation). As labour-intensive job-creation is the best approach to poverty alleviation, and as a huge 30% of the world's labour force is under- or unemployed, all taxes on labour should be repealed promptly. Firms must be encouraged to create more jobs and use less throughput. The social benefits of reducing unemployment are significant. The 'Cairo Consensus' (Cohen & Richards, 1994) amplifies these priorities.

The harsh truth is unpalatable world-wide: reproductive freedom is incompatible with population control — by definition. Paraphrasing J. S. Mill, one of the greatest champions of liberty who ever lived: First, to produce a child if that child does not have at least the ordinary chances of a decent existence is a crime against that child. Second, state laws which forbid marriage unless the couple can show that they have the means of supporting a family are legitimate powers of the state, and are not violations of liberty. Third, such laws prevent injury to society. In other words, it is not an infringement of individual liberty to prevent offspring from being born into lives of 'wretchedness', nor where they harm third parties or society in general. A recent manifestation of protecting the freedom of society from irresponsible individuals has developed in New York City, where TB victims who persistently refuse to take their (free) anti-TB pills are imprisoned or incarcerated in a locked hospital room until they are cured.

---

* and goodness knows how many 'loose' unmarried ones! — Ed.

‡ Here we feel it necessary to point out again that disciplined practice of due self-control and physical constraint could (and surely should!) constitute a major factor. — Ed.

Limiting Affluence

Since 1950, the world's human population has more than doubled in size, rising from 2.6 to 5.8 thousand millions. Further, an additional 2.5 thousand millions are expected by AD 2020. During this same period, world economic activity has risen five-fold. Despite such dramatic growth, however, far more individuals live in abject poverty today than at any other time in history, and income distribution has become increasingly inequitable. As human activities begin to outstrip the carrying capacity of biological support systems, we are led to ask, 'How much is enough?' — or even, how much can The Biosphere possibly support?

Examining consumption over the past four decades, one finds that *per caput* consumption of copper, energy, meat, steel, and timber, has doubled; *per caput* car ownership and cement consumption have quadrupled; *per caput* plastic use has quintupled; and *per caput* aluminium consumption has increased sevenfold. Distribution of these goods overwhelmingly favours industrial nations. Comparing affluence among particular nations, one discovers that a single US American on the average uses the same amount of energy as:

- 3 Japanese
- 6 Mexicans
- 13 Brazilians
- 14 Chinese
- 38 Indians
- 168 Bangladeshis
- 280 Nepalis
- 531 Ethiopians.

Indeed, while 260 million Americans spend US$5 thousand millions annually on special diets to lower calorie consumption, some 400 million people are so undernourished that they are likely to suffer stunted growth, mental retardation, or death (Daily & Ehrlich, 1992).

Maintaining high standards of living in the United States, Western Europe, Japan, Australia, parts of the Middle East, Singapore, and Hong Kong, requires enormous inputs of commodities — including energy, chemicals, metals, and paper — which cause or are the basis of extreme environmental damage. Data indicate that the residents of upper-income countries require the continuous production of from four to six hectares of land *per caput* to support their high-consumption life-styles. If the entire world population of 5.8 thousand million humans were to use land at this rate, the total land requirement would be over 26 thousand million hectares. However, the total land area of Earth is slightly more than 13 thousand million hectares, of which only 8.8 thousand million hectares is productive cropland, pasture, or forest. In short, the living standards prevalent in industrialized nations can hardly be called sustainable! Indeed, one is forced to ask, 'Cannot wealthy nations reduce their *per caput* consumption of resources, and lead the world's transition towards global environmental sustainability?'

Assessing *per caput* resource consumption among upper-income nations, one may be inclined to examine *per caput* gross national product (GNP) to ascertain levels of socio-economic welfare and the impact of economic activities upon national resources. Indeed, GNP statistics seek to account for the value of all final goods and services. Net National Product (NNP) does the same, but subtracts the depreciation of plant and equipment from overall output of goods and services.

Unfortunately, GNP figures are flawed measures of socio-economic and environmental health. First, traditional national income accounts *disregard the depreciation of natural capital*, such as soil erosion and destruction of forests by acidic precipitation, reduction of photosynthetic capacity and hydrologic cycles, and depletion of the stratospheric ozone shield (Kellenberg & Daly, 1994). While gross world product rose from 3.8 American trillion (or British billions, each of one million millions) dollars in 1950 to 18.7 such trillion dollars in 1991, the effects of environmental degradation and depletion of natural capital were not included in terms of economic costs. As a result, Brown (1993) noted that almost all countries were* 'practising the environmental equivalent of deficit financing in one form or another.' Present levels of consumption are currently being maintained through the exhaustion and dispersion of a one-time inheritance of natural capital — including topsoil, groundwater, and biological diversity, not to mention irreplaceable fossil fuels. Indeed, counting capital consumption as income is the worst of all accounting sins!

Second, economic sectors which contribute most to national income are frequently those which are most detrimental to local resources. Recent work on environmental accounting indicates that many environmentally benign activities contribute far less per unit of activity to GNP than environmentally harmful activities. For example, per unit consumption of mass transit systems contributes less to *per caput* GNP than automobile use. Similarly, train travel contributes less to *per caput* GNP than does airplane travel, and smaller families contribute less than larger families. In contrast, environmental damage and clean-up efforts contribute significantly to national income. Indeed, rectifying environmentally harmful

* and seemingly still are, *cf.* LESTER R.BROWN *et al.*'s subsequent editions of the Worldwatch Institute's admirable annual *State of the World* (*e.g.* Brown *et al.*, 1996). — Ed.

growth, such as nuclear and toxic clean-ups, is staggeringly expensive. Clean-up costs from the *Exxon Valdez* oil-spill, expected to exceed US $2.5 thousand millions, will be added to GNP, whereas the loss of tens of thousands of birds, sea mammals, and fish, will not be subtracted.

In spite of the dramatic economic growth which occurred from 1950 to 1990, one finds little correlation between affluence and income equality during that period. Whereas, in 1960, the richest 20% of the world's population had incomes 30 times as much as the poorest 20%, by 1990 they were receiving *60* times as much of the world's income. Indeed, traditional views of development which emphasize global income growth serve to exacerbate income inequalities while scarcely denting poverty.

How does this occur? Imagine that global incomes increase at an annual rate of 3%. In the first year, this translates into an increase of $653 *per caput* for the United States, but less than $12 for nations such as China, India, Bangladesh, and Nigeria. After ten years of 3% increases, *per caput* incomes in the US will have risen by $7,494 whereas *per caput* incomes in Ethiopia will have risen by only US $41. Thus, advocates of traditional views of development which place a priority upon global income-growth, actually serve to worsen the disparities of income distribution between individuals in rich and those in poor nations. When dealing with market competition for finite resources, income disparity becomes of critical importance. Indeed, relative income is more influential than absolute income in determining whether some individuals are excluded from access to resources.

From environmental and economic perspectives, we are concerned that the prescription to raise Northern income as a means to promote growth in the South fails to alleviate poverty, worsens income distribution within and among nations, and reduces environmental sustainability (Goodland & Daly, 1993a). Such prescriptions leave economic and social progress as mere 'spectator sports' for the poor of the developing world, which they neither participate in nor benefit from. Increased income-disparity between upper-income nations and low- and middle-income nations is likely to foment social stress as well as lead to an increase in environmental and economic refugees.

In recent decades, poverty and environmental degradation have been increasingly linked. The poor have not only suffered disproportionately from environmental damage, but have become a major cause of ecological stress themselves. Pushed onto fragile lands as a result of population growth and inequitable income distribution patterns, many of the developing world's poor have overexploited local resource-bases, 'sacrificing the future to salvage the present'.

Short-term strategies such as slash-and-burn agriculture, abbreviated fallow periods, harvests exceeding regeneration rates, depletion of topsoil, and deforestation, permit survival in the present but place enormous burdens upon future generations.

Genuine, long-term poverty alleviation, on the other hand, reduces pressure on the environment — through reduction in over-harvesting, overgrazing, or over-fishing — to meet short-run subsistence needs. By generating additional output and increasing the purchasing power of the poor, poverty *alleviation* widens domestic markets. Inclusive local markets offer greater potential for employment creation and increased self-reliance upon local resources to satisfy local needs.

Despite budget constraints, we believe that upper-income nations should see it to be in their own self-interests to invest in the South — to reduce inequality, to alleviate poverty, to protect and improve the global environment, and to avoid increasing the number of environmental refugees world-wide. Furthermore, development and transfer of up-to-date, efficient technologies should be accelerated to developing nations to process raw materials more efficiently than they do at present. Other priorities include renewable energy generation, increasing fuel efficiency, contraception methods, waste prevention, recycling means and methods, pollution prevention, sustainable and regenerative agriculture, and methods for reducing materials-use and energy-intensity in manufacturing (Goodland & Daly, 1993*b*).

IMPROVING TECHNOLOGY

At the current rate, world population is expected to double in the next 42 years, and given present levels of consumption, many ask, 'Can technological advances save us? Can we sufficiently lower *per caput* impacts through increased efficiency, such that reductions in living standards are unnecessary?' Ehrlich (1990) noted that, while technical progress will undoubtedly lead to efficiency improvements and resource substitutions, '... there is little justification for counting on technological miracles to accommodate the [American] billions more people soon to crowd the planet.'

The promise of present technology appears dismal. Western Europe's forest productivity is declining from sulphur dioxide emissions and those from fossil fuel-burning power-plants, factories, and automobiles. Seventy-five per cent of Europe's forests are now experiencing damaging levels of sulphur deposition.* In Bulgaria,

---

* Whereas it is in no sense a controlled scientific observation and certainly must not be cited as such, it is our strong impression that, at least as regards some forms of atmospheric pollution, trees may have some inherent capacity for getting used to it and even recovering to the extent of producing vigorous young growth over a few years' time-span, while different species evidently vary in their reactions, as do individuals according to their ecological state, etc. — Ed.

those living near heavy industrial complexes have an asthma rate that is nine times as high as that of people living elsewhere. On a global level, the World Health Organization (WHO) and the United Nations Environment Programme (UNEP) report that 625 million people are exposed to unhealthy levels of sulphur dioxide from fossil-fuel burning. Indeed, more than one thousand million people — nearly 20% of the Earth's population — are exposed to potentially health-damaging levels of air pollutants of various kinds.

What would it take to raise living standards in the South to the present levels of affluence in Northern nations? To answer this question remember first that, at the current rate, population is projected to double in the next 42 years, and realize that average *per caput* income in industrialized nations (US $18,330) is twenty-three times as great as that of people in low- and middle-income countries ($800). Thus, to raise Southern affluence to today's level of the North (holding impact and Northern affluence constant) implies improvements in technological efficiency by 2 (from a doubled population) x 23 = 46 times. However, historical improvement rates have never exceeded more than a tiny fraction of this rate!

This 46-fold increase must be achieved *via* resource efficiency, as opposed to increases in capital or labour efficiency. Historically, much of the increase in capital and labour efficiency has been at the expense of resource-use efficiency. In agriculture, for example, the increase in labour and capital productivity has required an enormous increase in complementary resource throughputs (*i.e.* energy, fertilizer, biocides, water, etc.), of which the individual productivity has subsequently declined. While technological improvements in agriculture, energy, transportation, and urban planning, may reduce the environmental damage caused by current systems, a 46-fold advance is simply not feasible. In short, it will be exceedingly difficult for poor countries even to approach the living standards of wealthy nations.

Increases in labour productivity have gone relatively too far. Unemployment is high (some 30%), rising, and seems to be ingrained in large parts of the world. Now, emphasis is needed on scarce natural resources' productivity — especially energy — rather than on abundant labour and capital. The burden of taxation must be shifted away from employment and on to natural resources, the new limiting-factor of production. 'Make it more profitable to lay off kilowatt hours, rather than people,' as Ernst von Weizsacker (1992) puts it. Shifting the tax-base away from income and on to throughput, would foster efficiency in energy and use of materials, and would help to internalize depletion and pollution's externalities (Weizsacker & Jesinghaus, 1992).

128   ROBERT J.A. GOODLAND & HERMAN E. DALY

A second means of approaching this issue is to examine energy-use as an imperfect surrogate for *per caput* impact (Daily & Ehrlich, 1992). In 1990, the wealthiest 20% of the world's population used an average of 7.5 kilowatts (kW) per person, for a total energy use of 9.0 terawatts (TW = 1 x 10^{12} watts). In contrast, the remaining 80% of the world's population used an average of 1 kW per person or 4.1 TW in aggregate. The total environmental impact of this energy-use was thus 13.1 TW.

Assuming that present population-growth will eventually be halted at 12 thousand million people, and that development succeeds in raising global *per caput* energy-use to 7.5 kW (approximately 4 kW below current US energy-use), the total energy-use would reach 90 TW. However, mounting environmental evidence indicates that present energy levels are unsustainable, and thus one needs little imagination to picture the environmental hazards of a seven-fold increase in energy use compared with that currently produced. Of course, should use of renewable energy (*e.g.* solar) suddenly burgeon, that would vastly reduce the impacts. Therefore, we believe that technological improvement, while always to be encouraged, will be far from sufficient to allow us to continue complacently to neglect the need to limit population and profligacy.

PRIORITIES FOR THE TWENTY-FIRST CENTURY

In the light of these findings, we suggest that the first means through which wealthy nations can assist poorer nations is by adopting the first oath of Hippocrates: '*non noli nocere* — first cease doing harm'. Curtailing demand for primary products harvested from fragile environs in tropical nations, halting export of toxic substances and harmful technologies (*e.g.* over-intensive monoculture farming), reducing emissions of 'greenhouse' gases, providing economic incentives for the preservation of biological diversity, and stopping military arms-sales to developing nations, are immediate suggestions. Likewise, acceleration of technological advances required to reduce the natural-resource content of given economic activities, and imposition of taxes on environmentally harmful throughputs (*i.e.* carbon emission, mineral severance taxes 'based on minerals leaving the mine', etc.), should all be supported.

In seeking to pursue a course towards environmental sustainability, high-income nations will need to adapt themselves far more than low- and middle-consuming nations — *see* Tables II, III, and IV.

Technological change and population stabilization cannot suffice to move the world towards an environmentally sustainable future. Instead, a reduction in *per caput* consumption in high-income na-

TABLE II

Means to Approach Population Stability.

1. Delay First Birth: delaying marriage and spacing births are tremendously effective in reducing family size.
2. Educate Girls: at least to level now enjoyed by boys.
3. Empower Women: Grameen-type mini-loans*; job creation; reduce gender discrimination.
4. Democratize Access: expand access to family planning choices from the rich to the poor.
5. Meet Unmet Demand: a huge 30% of women in some countries want family planning but have no access to it.
6. Premature Parenthood: campaigns against teenage pregnancy are essential.
7. Abortion: an abomination to many, family planning reduces abortion, and is surely preferable to infanticide & abandoned babies as regrettable alternatives. Legal abortion doesn't kill mothers.
8. Social Security and social safety-nets, including nutritional, maternal, reproductive, and child health.
9. Educate Men: males must act responsibly with self-control as partners and as fathers; training and raising consciousness are needed.

* This refers to Bangladesh's Grameen Bank which loans tiny amounts mainly to poor women.

TABLE III

Priorities for High-income Nations.

1. Transforming the life-style of consumerism in high-consuming nations into an ethos of environmental sustainability, or sufficiency for a stable population;
2. Internalizing environmental costs in energy prices and accelerating the transition to renewable energy sources;
3. Internalizing the costs of disposal of toxic and non-toxic wastes to the producer, and halting exports of such wastes to low-income nations (such measures provide incentives to minimize toxic-waste generation in industrialized nations); 'cradle-to-grave' pricing;
4. Accelerating technology development and transfers to low-consuming nations, such that they may leap-frog and consequently avoid environ-mentally-damaging stages of economic evolution; writing off unpayable loans held by debtor nations, particularly where liquidation of natural capital is required to finance loan repayments; increasing grants for low-income nations to reach global environmental standards (recent World Bank improvements in this respect are encouraging);
5. Broadening conventional cost–benefit analysis to internalize environmental costs;
6. Supporting the maintenance of biophysical infrastructures upon which all economic activity is built, and financing environmental investments as extended infrastructure investments; and
7. Focusing upon direct assistance to the poor in low-consuming nations rather than expecting general economic development to alleviate poverty.

TABLE IV

Priorities for Attaining Environmental Sustainability in Low-consuming Nations.

1. Accelerate the transition to population stability: democratize and encourage access to family planning;
2. Accelerate the transition to renewable energy;
3. Promote human-capital strengthening, with particular emphasis upon improving education, training, and employment creation, for girls and young women: job creation;
4. Poverty alleviation: empowerment of women, improved health-care for mothers, infants, and children, social safety-nets; and
5. Provide increased support for conservation and management of natural and other resources such as forests, croplands, and water.

tions and a decrease in environmental throughput are required. Maximizing efficiency and making full use of renewable energy would enable the current world population to live at approximately the level of Western Europeans in the mid-1970s. Such living standards allow for modest but comfortable homes, refrigeration of food, a moderate quantity of hot water, and access to public transportation. However, the high-consumption life-style standards that are prevalent most notably in the United States, with their larger homes, more numerous electrical gadgets, and (often solo) automobile-centred transportation modes, cannot be generalized. Indeed, even present levels of affluence in the rich nations are being financed through environmental exploitation and inequitable trading arrangements with developing nations, neither of which can be sustained for much longer.

It is neither ethical nor beneficial to the environment to expect low-income nations to arrest their development, which is currently highly correlated with throughput growth. Thus, in the transition towards environmental sustainability, high-income nations, which are responsible for a significant portion of present environmental damage and whose material well-being is able to sustain reductions in throughput growth, *must take the lead*. A major change in attitudes is needed in order to redefine development as improved human welfare, rather than GNP growth. Without such changes, we will continue to overshoot the carrying capacity of the Earth, impoverishing ourselves and future generations.

CONCLUSION

Unbridled consumerism in developed nations, marked by increased use of limited and often finite environmental resources, competes with escalating population-growth in less-developed nations, as the principal threat to environmental sustainability and the preventive of 'sustainable development'. We have examined the issue of sustainability through the $I = P \times A \times T$ identity, whereby the impact of any nation upon environmental sources and sinks is a product of its human population-numbers, their level of affluence, and the damage done by the technologies which support that affluence, and have found this equation to be useful for guidance.

Population stabilization is essential in the transition towards hoped-for sustainability. Likewise, reduced consumption of materials and energy, poverty alleviation, and improved management of environmental resources, are crucial for this transition, in which technology transfers to low-income nations will be vital in reducing environment-detracting throughput. Priorities for high- and low-income nations include changes in public perceptions about the concept of development if we are to avoid overshooting the carrying capacity of Planet Earth.

REFERENCES, AND FURTHER READING (especially items ending with a '+')

AGARWAL, A. & NARAIN, S. (1991). Global warming in an unequal world. *International Journal of Sustainable Development*, 1(1), pp. 98–104.

BROWN, L.R. (1993). A new era unfolds. Pp. 11–22 in *Worldwatch Institute — State of the World 1993*. W.W. Norton, New York, NY, USA: xvi + 256 pp.

BROWN, L.R., *et al.* (1996). State of the World 1996. W.W. Norton, New York, NY, USA: xviii + 249 pp.

CLARK, C. (1991). Economic biases against sustainable development. Pp. 319–30 in *Ecological Economics: The Science and Management of Sustainability* (Ed. R. COSTANZA). Columbia University Press, New York, NY, USA: 544 pp.

COHEN, S. & RICHARDS, C. (1994). The Cairo Consensus: population, development and women. *Family Planning Perspectives*, 26(6), pp. 272–7.

DAILY, G.C. & EHRLICH, P.R. (1992). Population, sustainability, and Earth's carrying capacity. *BioScience*, 42(10), pp. 761–71.

DALY, H.E. (1972). In defense of a steady-state economy. *Amer. Journ. Agr. Economics*, 54(4), pp. 945–54.

DALY, H.E. (1974). On the economics of the steady state. *Amer. Econ. Rev.* (May) pp. 15–21.

DALY, H.E. (1989). *Sustainable Development: From Concept and Theory Towards Operational Principles.* Paper presented to Hoover Institution Conference. [Not available for checking.]

DALY, H.E. (1990). Carrying capacity as a tool of development policy: the Ecuadoran Amazon and the Paraguayan Chaco. *Ecological Economics*, 2, pp. 187–95.

DALY, H.E. (1992). Allocation, distribution and scale: towards an economy that is efficient, just and sustainable. *Ecological Economics*, 6(3), pp. 185–93.

DALY, H.E. & COBB, J. (1994). *For the Common Good* (2nd edn). Beacon Press, Boston, Massachusetts, USA: viii + 355 pp.

DURNING, A. (1989). *Poverty and the Environment: Reversing the Downward Spiral.* (Worldwatch Paper 92.) Worldwatch Institute, Washington, DC, USA: 50 pp.

EHRLICH, P.R. (1990). *The Population Explosion.* Simon & Schuster, New York, NY, USA: 320 pp.

132   ROBERT J.A. GOODLAND & HERMAN E. DALY

EHRLICH, P.R. (1994).

EHRLICH, P.R. & HOLDREN, J.P. (1974). Impact of population growth. *Science*, **171**, pp. 1212–7.

EL SERAFY, S., *see* SERAFY, S. EL.

GOODLAND, R. [J.A.] (1991). The case that the world has reached limits. *Population and Environment* **13**(3), pp. 167–82.

GOODLAND, R. & DALY, H. [E] (1993a). Why Northern income growth is not the solution to Southern poverty. *Ecological Economics*, **8**, pp. 85–101.

GOODLAND, R. & DALY, H. [E] (1993b). *Poverty Alleviation is Essential for Environmental Sustainability.* (Environment Department Divisional Working Paper No. 1993-42.) The World Bank, Washington, DC, USA: 34 pp.

GOODLAND, R., DALY, H.E. & SERAFY, S. EL (1992). *Population, Technology and Lifestyle: the Transition to Sustainability.* Island Press, Washington, DC, USA: 185 pp.

GOODLAND, R.J.A., DALY, H. & SERAFY, S. EL (1993). The urgent need for rapid transition to global environmental sustainability. *Environmental Conservation*, **20**(4), pp. 297–309, fig. and 7 tables.

HARDIN, G. (1993). *Living within Limits.* Oxford University Press, New York, NY, USA: 288 pp.

HICKS, SIR J. (1946). *Value and Capital.* Clarendon Press, Oxford, England, UK: 446 pp.

KELLENBERG, J. & DALY, H.E. (1994). *Counting User Cost in Evaluating Projects involving Depletion of Natural Capital: World Bank Best Practice and Beyond.* World Bank Environment Division Paper, World Bank, Washington, DC, USA: 67 pp.

LAPPE, F.M. & SCHURMAN, R. (1990). *Taking Population Seriously.* Institute for Food and Development Policy, San Francisco, California, USA: 90 pp.

MEADOWS, D.H., MEADOWS, D. & RANDERS, J. (1972). *Limits to Growth.* Universe Books, New York, NY, USA: 205 pp.

MEADOWS, D.H., MEADOWS, D. & RANDERS, J. (1992). *Beyond the Limits.* Chelsea Green Publications, Post Mills, Vermont, USA: xix + 300 pp., illustr.

PINSTRUP-ANDERSEN, P. (1993). Perception on food supply in developing countries. *Outlook on Agriculture*, **22**(4), pp. 225–32.

POLUNIN, N. (Ed.) (1974–95). *Environmental Conservation* (quarterly journal devoted to global survival) published for the Foundation for Environmental Conservation. From Vol. 23, 1996, Ed N.V.C. Polunin and published by Cambridge University Press. +

POLUNIN, N. & BURNETT, Sir J. (Ed.) (1993). *Surviving With The Biosphere: Proceedings of the Fourth International Conference on Environmental Future (4th ICEF), held in Budapest, Hungary, during 22–27 April 1990.* Edinburgh University Press (for the Foundation for Environmental Conservation), Edinburgh, Scotland, UK: xxii + 572 pp., illustr. +

ROYAL SOCIETY OF LONDON & US NATIONAL ACADEMY OF SCIENCES (1992). *Population Growth, Resource Consumption, and a Sustainable World.* London & Washington, Royal Society of London and US National Academy of Sciences: [not available for checking].

SERAFY, S. EL (1991). The environment as capital. Pp. 168–75 in *Ecological Economics* (Ed. R. COSTANZA). Columbia University Press, New York, NY, USA: 544 pp.

SERAFY, S. EL (1993). *Country Macroeconomic Work and Natural Resources.* The World Bank, Washington, DC, USA: Environment Working Paper 58, 50 pp.

SERAGELDIN, I., DALY, H.E. & GOODLAND, R.J.A. (1995). The concept of sustainability in W. van DIEREN (Ed.) *Sustainability and National Accounts.* Amsterdam, IMSA (for The Club of Rome) (ms. 28 pp.).

UNITED NATIONS DEVELOPMENT PROGRAMME (1993). *Human Development Report 1993.* Oxford University Press, New York, NY, USA: 114 pp.

WEIZSACKER, E.U. VON (1992). *Why the North Must Act First.* International Academy of the Environment, Geneva: 31 pp.

WEIZSACKER, E.U. VON & JESINGHAUS, J. (1992). *Ecological Tax Reform: a Policy Proposal for Sustainable Development.* Zed Books, London, England, UK: 92 pp.

WORLD BANK (1992). *World Development Report 1992: Development and the Environment.* Oxford University Press, Washington, DC, USA: xii + 308 pp., illustr. +

7

Health of People, Health of Planet

by

Morris Schaefer, DPH (Syracuse)

Professor (Emeritus) of Health Policy and Administration,
University of North Carolina School of Public Health,
Chapel Hill, North Carolina 27599, USA;
Public Health Management Consultant,
1817 Hobbs Road,
Greensboro,
North Carolina 27410-3925,
USA,

&

Wilfried Kreisel, PhD (Heidelberg)
Executive Director, Division of Environmental Health,
World Health Organization,
Avenue Appia 20,
1211 Geneva 27, Switzerland

Introduction*

It is practically axiomatic that the health of humans is determined by their genetic heritage and how they interact with the environments in which they live. These environments provide sustenance and protection, but also present hazards to health and ultimately to life.

Even in primitive communities, human environments are complex, being always made up of many interrelated physical and social elements. Social elements impact heavily on the physical environment, as human activities continuously alter natural conditions.

Among these social elements, demographic factors are powerful determinants of the state of the environment and, thereby, the state of human health. Both the environment and human health are now endangered on a global scale, and demographic factors are crucial in this crisis.

---

* The present chapter is not referenced in detail, but ends with a list of Supporting Documents that cover points made in this survey but do not necessarily reflect the opinions of its Authors. — Ed.

134

The concept of health is defined in the Constitution of the World Health Organization (WHO) as 'a complete state of physical, mental, and social, well-being, and not merely the absence of diseases'. In practice, health measures are usually aimed at averting, preventing, and curing, diseases and impairments — mainly those of the human body. To varying degrees, communities are also concerned with mental diseases, and most seek to control the social 'diseases'[†], crime, violence, destitution, and alienation.

Most bodily diseases and disabilities are caused, directly or indirectly, by environmental factors — even including the role of genetic traits and physiological processes, such as ageing and the vulnerabilities of infancy and pregnancy. In fact, some of these physiological processes may be accelerated, retarded, or intensified, by environmental conditions, and the environment may trigger specific diseases when people are physiologically and psychologically vulnerable. The state of one's environment is integrally intermeshed with the state of one's health.

Generally, disease results from *exposure* to a *pathogen* in the environment, against which a person's or group's *defence* is inadequate. Such *pathogens* may be biological (mainly microorganisms), chemical (natural toxins, gases, dusts, cigarette smoke, manufactured compounds, or drugs), or physical (hazards of mechanical trauma, temperature extremes, etc.); many of these pathogens can be socially controlled. *Exposures* to, and dangers from, pathogens may also be increased by such social practices as drug-use, prostitution, war, and risk-taking; or they may be lessened through social provisions for occupational safety, community hygiene, or control of pathogenic sources, vectors, and hazards.

Bodily defences against disease include one's general state of nutrition and fitness, specific natural or induced immunities, and behaviours that avoid or reduce exposures. Again, one's bodily defences may be enhanced by social arrangements to provide adequate and safe food, water, and shelter, immunization and other prophylaxis against disease, and education in healthy personal practices; in the absence of such community provisions, the likelihood of disease and impairment is increased.

PERCEPTIONS OF ENVIRONMENT

We each of us perceive our respective environments as a set of living conditions — a whole situation — more or less shared with those around us. However, to understand contemporary threats to health — and the possibilities for improved health — requires that

---
† *i.e.* sexually transmitted diseases.

we analyse the environments in which humans live, to understand their 'anatomy and physiology'. The major analytical distinction is between the *physical* and *social* environments.

Physical Environment

In the physical environment we distinguish between natural conditions and human-made structures and technologies (the 'built environment' or technosphere). The *natural* elements of air, water, and soils, are the basis for life itself. The plant and animal species thus supported have interdependent relationships (sometimes antagonistic) among species and with The Biosphere as a whole. Notable in The Biosphere is a widespread dependence on generally stable climatic conditions and on the basic life-process of carbon dioxide/oxygen conversion through plant photosynthesis. The equilibria of natural elements and biota were built up over aeons in The Biosphere, and involve different climatic and geographical zones, as well as local ecosystems and more regional ecocomplexes. When such equilibria are sufficiently disturbed, the capacity of ecosystems to support life is impaired, with direct and indirect effects on human health. 'Disturbances' take many forms, for example:

— Destruction of forest cover impairs biomass buildup from photo-synthesis, leads to a loss of soils through erosion and of species through disruption of their habitat, and may result, in the long term, in decreased agricultural productivity or even desertific-ation.
— Biological and chemical pollution of waters endangers human health — directly by infections and poisonings, and indirectly by weakening or destroying species at the base of the food-chain, including species that are used for human food and/or medic-ation.
— Chemical pollution of the atmosphere has the *direct* health-effects of inducing and intensifying pulmonary and cardio-vas-cular diseases and of increasing skin cancers through thinning the stratospheric ozone shield and consequent reduced filtering-off of ultraviolet sunlight. It also has the many *indirect* health-effects which include weakening and destroying forest, water, and marine, resources (for example, through 'acid rain' and excess ultraviolet light-rays), and inducing local climatic changes (such as temperature inversions) against which people may not be adequately protected.

The *human-made* component of the physical environment in-cludes tangible structures and technologies. *Structures* are represented by houses, farms, dams, factories, vehicles and other durable products, road networks, and systems for the generation and transmission of energy. Most such structures include constituent structures: a house, for example, has structures for heating, cooking, water-supply, and waste disposal. Each Man-made structure has a dual relationship with Nature: first, its existence and operations modify natural conditions by means of the space which it occupies, the resources that are consumed in it, and/or the wastes which are discharged from it; second, it is composed of materials that are drawn from Nature at some stage of its production, and many structures exist to convert natural resources into other forms of goods.

Technologies in the physical environment are established practices for energy and material conversion, including power-generation (from fossil and nuclear fuels, water movement, etc.), energy-use (as in motor vehicles, or for cooking and heating), manufacturing, agriculture, house-building, and waste management. Technologies usually consist of methods and equipment. The methods may be traditional or rationalized, and the equipment may be as sophisticated as an automated factory or as simple as a fire-pit or smoke-hole.

Technological capabilities are unequally distributed among countries, those in the industrialized world being generally far more advanced than those in the less-developed world. Within countries, technologies are often unequally distributed among districts and according to people's economic roles. Social variables are important in the distribution of technologies; among them are educational and scientific levels, degree of institutionalization, constraints of culture, patterns of economic organization, and legal arrangements for the ownership of intellectual properties.

Social Environment

Because it embodies human choices, actions, and behaviours — and their results — the social environment is a major determinant of the state of the physical environment. Human activities that impact on the physical environment have health consequences for both good and bad; in addition, the social environment itself can promote health or generate health-impairing pathogens.

The social environment consists mainly of structured relationships, which are highly stable in traditional communities but more 'dynamic' in modern and transitional communities. The obvious relationships in social environments are those among human beings; but the social environment also includes human relationships with things (both natural and Man-made) and with many processes —

educational, economic, cultural–religious, technological and, through governments etc., political and legal as well.

Human relationships are experienced through families and communities. People's community relationships are numerous, especially amid the complexities of modern urban settings. Individuals and their families always belong (if only by assignment) to geopolitical communities at the level of the settlement, and often to neighbourhood communities in larger settlements. Geopolitically, they also relate to states and provinces, nations, and associations of nations, although sometimes in only a hypothetical sense.

The word 'community' is most often used to denote people's relationship to a local settlement, but people also belong to religious, ethnic, occupational, partisan, cultural, linguistic, economic, generational, or gender, communities, many of which are interlinked, but some of which may be separate and in conflict with one another. Thus, membership in communities generally provides psychological, economic, and spiritual, supports; but community affiliations may also be a source of stress among individuals and families — especially when they exist in opposition to one another, which is often the case in transitional societies.

The values that communities share will influence their members' individual and collective choices and behaviours. This holds true for affiliations that one is born into and grows up in, and also for affiliations that one chooses because they reflect one's values and concerns. These community influences on behaviour are of cardinal concern in the history, and for the future, of health and the environment.

Community influences and values strongly condition interpersonal behaviours, such as those involved in human reproduction, relationships within the family, the role of one's peers (*i.e.* numbers of groups formed around age, gender, ethnicity, or economic, situations), the character of any criminal activity, and sexual practices within and outside the family. Each of these behaviours or social practices has important health implications with respect to the frequency of maternal, infant, and child, diseases and impairments, sexually transmitted diseases, domestic violence and the abuse or neglect of children, traumas from accidents or crime, abuse of alcohol or addictive drugs including tobacco, and utilization of health, medical, educational, and other, social services.

Social values also condition community choices of basic provisions for health. This is most clearly seen in such public-policy areas as the provision of sanitary infrastructures (especially those that impinge on water-related diseases, including diarrhoea, cholera, and malaria), adequate shelter, transportation (in its engineering, regulatory, and pollution, aspects), food safety and distribution, health care, schools and adult education services, recreation, and measures to reduce or offset the effects of poverty. Studies in social epidemiology testify that the priorities accorded to meeting such basic requirements, and the degree of equity in their distribution among areas and social groups, are major determinants of health outcomes and the incidence of well-being or misery within populations.

ROLE OF SOCIO-ECONOMIC DEVELOPMENT

Of critical importance for the future of Planet Earth and human health are the ways that human communities (comprising the social environment) use technology and natural resources, thereby modifying the physical environment in the process that is generally denoted as socio-economic development. In development, people exploit Nature through economic activities in agriculture, lumbering, manufacturing, fishing, and extraction of minerals and energy-yielding fuels; this exploitation generally increases as populations grow and developmental activities themselves accelerate.

Nature is also altered by the social activities of settlement, recreation, and the maintenance of persons and communities. All these activities act to modify lands, waters, other living species, and the atmosphere; further modifications occur from the need to accept the wastes of consumption and production. Many of these interactions — including bodily wastes of the social and physical environments — have adverse effects on health, both short-term and long-term. For example, heavily polluted air in some large cities increases pulmonary and cardio-vascular disease-risks, and poor land-use and overfishing have already reduced food supplies in some regions. In the long term, exceeding the bearing capacities of ecosystems, of ecocomplexes, and ultimately of The Biosphere, could have catastrophic implications for human health, in the end affecting the very viability of other forms of life. Already quite numerous species of plants and animals are extinct or approaching extinction.

When humans choose to exploit often-changing technological possibilities, whether or not through organized development programmes, environmental impacts are invariably felt. For example, shifting from organic to chemical technologies in agriculture not only affects food production and safety, but also modifies the conditions of lands, waters, and aquatic biota, in the surrounding ecosystems. Health effects may be experienced not only in agricultural *locales*, but also in far-away urban communities. Likewise, the building of highway systems (or regional dams for irrigation and energy production) not only modifies the use of the often-vast lands that are taken for the roads, dams, and impounded lakes themselves, but results in even far greater and wider environmental impacts by enhancing possibilities for industrialization, new and enlarged settlements, commercial activities, irrigation farming, forestry, and mining. At the same time, such developments can have drastic negative side-effects in the destruction of species and their habitats, while resulting in other biotic disruptions such as the alteration of cultures, the dislocation of human communities, and the pollution and modification of lands, waters, and the atmosphere. Those negative effects may have transnational and even global impacts on the natural environment and human health.

Three leading variables in socio-economic development are human population, industrializing technologies, and development policies. Small and dispersed *populations* impose relatively little stress on the natural environment, and their ecosystems' capacity to provide sustenance and absorb wastes is virtually unimpaired. But when populations become large and are densely concentrated, especially in immense cities, stresses increase; resources are more rapidly depleted, and a large burden of wastes pollutes and otherwise modifies the natural environmental base for human health. Rapid growth of urban populations often presents an additional set of health-hazards, especially when that growth exceeds the community's capacity to provide adequate shelter, excreta-disposal, safe food and water, and social services — including education and medical care.

SUSTAINABLE DEVELOPMENT*

When human populations use *technologies* that make intensive use of energy and materials, ecosystem degradation is inevitable and commonly accelerated. The historical dominance of such technologies in the industrialized countries has resulted in disproportionate resource-depletion and waste-generation by their high-consumption economies. In so far as developing countries follow the same technological course in their socio-economic development, their populations' large size (with their preponderance of younger ages and higher growth-rates) would be likely to impose such stresses on

*Defined as 'development that meets the needs of the present without compromising the ability of future generations to meet their own needs', and hence taking care of all concomitant environmental/ecological situations. Because, in the minds of many, 'development' conjures up visions of structured aggrandizement, we are gratified to note the increasing modern tendency (which we practised for years in editing *Environmental Conservation*) to precede this phrase with 'environmentally' or 'ecologically'. — Ed.

the natural environment as to bring into question the planet's capacity to support such superabundance of human life.

Development policies — and their planning, adoption, and implementation — have traditionally concentrated on the application of technologies, often of increasing sophistication, to the conversion of natural resources into goods for human use. The political motivations have been varied, according to historical and current conditions. Most development is declared to be driven by needs or desires to improve living conditions — by meeting people's basic and not-so-basic needs. Yet the making of sound development policies must always confront a dilemma. For on one hand, development is the key to overcoming poverty for hundreds of millions of Earth's human inhabitants, and growing economies are widely believed to be necessary to sustain the scientific and technical infrastructure which is needed to solve emerging problems and raise living standards. Development's effects on well-being are positive when it equitably improves the availability of goods (notably food, safe water, and shelter) that are basic to human life and health, and equitably provides for protection against, and treatment of, disease and disability.

On the other hand, economic development almost always re- quires changes in land- and water-use, and it usually entails raising levels of energy-use and waste-generation. In addition, even if improving social wealth that is equitably distributed improves conditions for the promotion of health, modern development may increase health-hazards through pollution, poorly controlled applications of technology (*e.g.* inadequate protection of workers against adverse exposures to chemicals or physical hazards), and increases of a variety of stresses in changed living-situations. This dilemma is inescapable; it can be solved by merely finding a balance between long- and short-term interests and limiting unnecessary environmental deterioration. Therefore, how development is carried out has become a pivotal problem for human health and survival. The issue now is the degree to which socio-economic development can be made sustainable. (The concept of *sustainable development**** was enunciated by the World Commission on Environment and Development (1987); the 1992 United Nations Conference on Environment and Development, or 'Earth Summit', sought to implement that concept in political, technical, and administrative, terms.)

'Non-sustainable development' consists of changes in Nature that are, in practical and political terms, irreversible (as in the creation of large Man-made lakes) or irreplaceable — as in the extinction of

---

* *See* footnote in preceding page. — Ed.

species and the mining of fossil fuels and 'old' forest resources, whether in tropical, temperate, or even sub-polar, zones. 'Non-sustainable development' also occurs when destructive practices continue to the point where the recuperative capacities of The Biosphere or regional ecosystems or ecocomplexes are exceeded.

Non-sustainability in the short or medium term has health consequences for people now living, as evidenced in changes in communicable disease-incidence associated with some water-resource developments, traumas resulting from building shanty-towns on unstable slopes, decreased food supplies from improper use of land- and water-resources, and increases in diseases related to environmental pollution. In the long term, the projected consequences of non-sustainability could vastly increase disease-burdens, social and political instability, and hunger; at the extreme, the survival of the human species could come into serious question.

In the face of these trends, the sustainable development approach proposes that, in meeting human needs, demands on the planet's resources be stabilized, with reduction when necessary, and that consumed resources be replaced whenever feasible. Behind these general statements regarding sustainable development, an enormous and complex array of factors and issues — demographic, economic and financial, political and diplomatic, cultural and ideological, technological, and legal — come into play and require resolution and modification of many aspects of current practices. The significance and interplay of these factors and issues can be illustrated in considering world-wide trends in urbanization, which themselves have major implications for human health.

URBAN ENVIRONMENTS AND HUMAN HEALTH

Cities* have historically been the sites of, and driving forces behind, social and economic development. Whatever the character of a culture's adopted technology, its highest and most intense manifestations are found in large cities. The city can be thought of as a Man-made physical environment and, by definition, also a social environment.

Compared with rural environments, cities are characterized by high concentrations of people and structures, an advanced level of task-specialization, and relative geographical compactness as

---

* Here meaning large towns or major urban agglomerations, not limited to the traditional British sense of towns so designated by charter and containing a cathedral, or to the United States' municipal State-chartered corporations occupying a definite area — regardless of human population that in some cases of which we know is only a very few hundred people. — Ed.

measured by population densities. Because they are the prime *locales* for pursuing economic development, cities generally possess higher levels of wealth — at least, social or aggregated wealth — than rural areas. Thus most cities are important generators of wealth, serving as the principal and relatively efficient seats of industry and commerce, higher education and research, and cultural diversity. At the cutting-edge of development, cities account for generating most of the wealth in most countries; by recent World Bank calculations, the 34% of the global population living in cities produces more than 60% of gross domestic product (GDP).

Health in Cities

Historically, also, cities have been at the cutting-edge of developments in community health. Traces of municipal sanitary infrastructures have been found in the ruins of ancient cities of the Indus and Nile river valley cultures that flourished some four millennia ago; key public-health concepts developed when the walled cities in Europe's medieval period felt the brunt of devastating epidemics of disease (as have some contemporary cities in the developing world), and the first significant advance in environmental public-health controls was organized in the streets and slums of that continent's 19th-century industrial cities.

The relationship between urban environments and human health derives from the very nature of the city itself. On one hand, the concentration and proximity of people in those environments create conditions of disease communicability and may increase exposures to various biological, physical, chemical, and social, pathogens, some of which are peculiar to urban environments. On the other hand, it is people's residential and occupational proximity that makes feasible the provision of sanitary and other protective services; also, the relatively greater social wealth of cities makes such protections affordable, at least hypothetically and in comparison with most rural areas in developing countries.

At this juncture, expanding health-risks in the rapidly-growing cities of developing countries require the greatest attention. Risks of communicable diseases are increased by low levels of immunization in populations that are expanding by in-migration of rural dwellers. In many of these cities, environmental barriers to protect the local populace against biological pathogens are weak; provisions for housing, water-supply, excreta- and garbage-disposal, drainage of surface waters, hygienic practices, and food sanitation, are all likely to be inadequate, whether because of policies or because municipal authorities and financial resources are overwhelmed by the rate of population increase and concomitant demands.

Exposures to pathogens are increased when people live in crowded conditions and disease vectors are not controlled*; also, immigrants may introduce pathogens to which the established population is not resistant. Apart from the adequacy of community provisions to deal with growing populations, cultural factors may be implicated as, for example, when migrants continue traditional practices — such as collecting water from surface sources, or building dwellings with sanitary provisions unsuited to crowded urban conditions. Risks increase also when substantial numbers of people are inadequately educated in hygienic and other health-protection practices.

Much of the burden of chronic disease and traumas (especially poisonings, burns, and injuries) is associated with urban environmental factors, life-styles, inadequate early detection and treatment of disease, and poor health-education. Structural hazards in shelter, transportation, and occupation, are a major problem: buildings, equipment, and furnishings, may not have adequate safety features or may be built with or made from unsafe materials; no provision may be made to escape from fires, explosions, and collapses; and work stations in industries and homes may be poorly engineered.

Poor siting of settlements in relation to industries and waste-dumps — often the situation in which peripheral squatter families find themselves — increases people's exposures to concentrated pollution of air, water, and soil, as well as to direct contact with hazardous substances and germs. Exposures to acute and chronic illnesses rise also with the increased use of toxic and caustic chemicals in industries and homes, not all of the effects of which are known; accidents in manufacturing, storage, and transportation, increasingly give rise to emergencies involving hazardous materials.

Urban air-pollution involves many health-hazards from numerous sources, and control efforts involve severe technical and economic difficulties. Among the multiple health-effects are the toxic and degenerative effects of lead from vehicular emission (rising with traffic congestion), high levels of dusts and sulphur from energy generation, and industrial emissions that contribute to chronic pulmonary diseases, also the acute effects of smog, and the immediate and ecological impacts of 'acid rain'. Combustion within homes that have inadequately-vented heating and cooking devices, especially when coal and biomass fuels are used, puts hundreds of millions of people at risk in the developing countries.

Clearly, both health-risks and health-protection are strongly affected by the character of the technologies that are applied in the urban environment. Although agriculture has been (and is continuingly being) industrialized in many countries, industrialization is more often concentrated in the manufacturing, transportation, and commercial, activities of cities. Much as when Europe was indus-

---

* As they are now, rigorously and most mercifully, of the long-dread disease Smallpox — *see* footnote on page 156. — Ed.

trializing in the 18th and 19th centuries, developing countries have
tended to apply advanced technologies more rapidly to economic
production activities than to social provisions for sanitation, safe
energy-use, and adequate domestic living conditions for the great
mass of people. In addition, inappropriate policies and property laws
have further depressed the living conditions of economically
marginal and sub-marginal urban dwellers.

The health consequences of technology depend in part on the
interdependence between technology, space, and population density.
Urban communities must actively manage biological, chemical, and
physical, wastes — including emissions into air — because the ratio
of wastes generated in relation to the available disposal-space is
much higher in cities than in rural areas. Urban population concentration in itself magnifies the health impacts of natural and Man-
made disasters, just as it magnifies the effects of socially-based
breakdowns in water supply and sanitation, inadequate vector control, and communicable disease outbreaks.

Urban situations are especially vulnerable when applications of
industrial technology outrun the community's capacity for health
protection through regulation and control of environmental hazards
and exposures, including needed safeguards in the workplace and
residential environments; without such controls, people's risks of
disease and trauma increase. Again, the urban poor are usually the
most at risk; they are usually the most exposed to poorly-managed
hazardous wastes, occupational and traffic hazards, and air pollution
from manufacturing and traffic sources.

Urban living may induce pathologies that become rooted in the
social environment itself. The cultural environments of fast-growing
cities are notoriously unstable, because both the city and many of its
people are in transition. Both are moving from traditional social
relations and accustomed settings to new forms of association and
functioning; despite congestion and crowding, some people find
themselves living in family or individual isolation. In many cities in
developing countries, mass-communication technologies, which are
increasingly national and international in origin, introduce different
value-sets and make a significant contribution to increasing the
stresses of cultural instability. Psychological health is hurt by social
value-sets that condone the occupational and sexual exploitation of
women and children, the persistence of poverty and homelessness,
and the institutionalized insecurity of aliens and the poor. When
people feel themselves rootless, powerless, or out of touch with firm
moral values, their vulnerability to socially-based pathology can be
expected to increase, taking such forms as emotional instability,
social deviance, depression and even suicide, criminal and domestic
violence, drug dependence, and sexually-transmitted diseases.

Qualitative and Quantitative Factors in Urban Health

The health and social effects of such *qualitative* factors are heightened by the *quantitative* factor of rapid urban population growth. While global population growth itself adds to the drain on natural resources and the waste-burden imposed on the natural environment, urban population growth adds disproportionately to the problem. The rate of urban population growth over the last half-century has exceeded general population growth. Although about 40% of this urban growth has been attributed to migration from rural areas (with negative effects on the demographic and economic resources of rural communities), the larger share has come from natural increase; that is, births to people already established in the city.

By far the greatest increases have been in developing countries, where the urban growth-rates are 2–3 times those experienced in the past in the now-industrialized countries. The populations of cities in developing countries are widely expected to increase even more in the next half-century, rising to double the present urban populations of industrialized countries by AD 2000, and to four times those populations by AD 2025. Such growth will have enormous impacts on the physical environment.

Special concerns attach to the projected growth in the number of 'megacities' (with from 5 million to 30 million inhabitants), in some of which extreme conditions of air pollution and regional climatic changes are already being experienced, *e.g.* Mexico City. While there was only one Third World city of more than 5 millions in 1950, the number had risen to 11 by 1970 and may reach 35 by AD 2000. Yet, despite the growth of megacities, most 21st-century urban dwellers will live in cities of fewer than 5 millions' population, and while health and environment problems are likely to be more intense in megacities, such problems will be more extensive in the remaining cities.

Recent and projected urban growth in developing countries is of great importance to the state of their health and environments. Most developing countries already face a major backlog in meeting the basic health-needs of large numbers of their people, have less social wealth for providing environmental health protection than do industrialized countries, and have weaker environmental control systems and scientific infrastructures. In city after city, the population growth has come about too rapidly for their social systems to absorb: for example, municipal authorities have often been overwhelmed by the size and speed of this growth and its social requirements of sanitation, housing, education, and medical care; land-tenure laws are either inadequate or so bitterly contested as to leave large, impoverished populations with no choice of dwelling except in squatter settlements and shanty-towns, under inadequate or even precarious living conditions; and ever-growing numbers of people are left to live in extreme poverty, at high levels of physical and social risk.

In situations where urban birth-rates exceed rates of economic growth, as was the case during the 1980s, the inevitable result is the further extension of poverty, which is the social condition most clearly related to poor health. Many cities in developing countries support a large 'informal economy' that is not only marked by low wages and a lack of necessary controls over working conditions, but also has the result that the poor pay more for — and have less of — such necessities as food and water for drinking and hygiene. Large numbers of poor people add to urban environmental and health stresses by the acts necessitated by their very struggle to survive; such stresses increase when urban municipal authorities are unable to manage adequately the necessary domestic and sanitary wastes' disposal, drainage of standing waters, vector control, and growing levels of pollution.

Even though population growth has been largely stabilized in cities in industrialized countries, those cities severely stress the natural environment — often well out of proportion to their popu-lations. These stresses derive mainly from high levels of consump-tion — based on technologies that are advanced, energy- and material-intensive, and dynamic to the point that some technology applications often outrun established controls. Other aspects of the social environment add stresses (with adverse health-effects) in many Western cities. These stresses take many forms, such as ageing sanitary, housing, and transportation, infrastructures; movement of the more affluent inhabitants to suburbs, leaving core urban areas to decay; and cultural shifts that have economic and behavioural effects. Many Western cities experience employment shifts with economic cycles and technological changes, sometimes resulting in increases in the numbers of people living in conditions of poverty and homelessness. In some cases, international migration adds to economic and social stresses in the urban environment.

Urban Settlements and Human Ecology

The full environmental impacts of rapid urbanization cannot be counted without considering urbanization's effects on the hinter-lands on which cities depend. Rising urban demands for energy, foodstuffs, water sources, and raw materials, remake rural eco-nomies and rural landscapes. Road networks are developed to speed the movement of goods and to reach towards 'new lands'. Forests shrink as demands for wood and cropland increase. Agricultural production targets change from subsistence- to cash-cropping, and techniques shift to mechanization and 'chemicalization' — often with ill-effects on soil reserves, water quality, and aquatic life.

Cities have always drawn from rural zones a large part of the resources which they process and use to sustain themselves, including the key resources of human labour that is attracted to cities and the conversion of agricultural land to residential and industrial uses. In the past, a city's hinterland was usually the rural areas proximal to it. But from antiquity onwards, some cities were located to enable their engagement in maritime and overland trade, which extended their resource fields (and market areas) well beyond their immediate settings. Urban-based commercial and manufacturing interests provided much of the impetus for Europe's mercantile and colonial expansion from the 16th century onwards. Nowadays, as more societies adopt advanced technologies, cities and national economies draw massively on distant resources, as represented in international trading of goods that range from petroleum to fresh vegetables, fruits, and flowers.

As with the extraction of resources, the space-dimension has grown increasingly important in the disposition of urban wastes. In earlier periods, the city-generated wastes were usually disposed of nearby: that is, they were released into the atmosphere, dumped onto land, or discharged into water — either rivers or streams or, in the case of coastal cities, offshore. As cities and metropolitan conurbations have grown and spread over larger and larger areas, forming urban belts and corridors, disposal on land has become less feasible, if only because 'near by' is often at a considerable distance. Discharges into water produce adverse ecological effects (such as algal blooms, fish-kills, and 'dead' lakes) and such health effects as epidemics of water-borne diseases. Airborne waste emissions not only pollute the immediate environment (with measurable adverse health-effects) but have also produced such 'distant' ecological threats as 'acid rain' and stratospheric ozone destruction. Difficulties have been compounded as the consumption demands of growing populations produce ever-larger volumes of waste, at the same time that some technological innovations produce new types of hazardous wastes, including the extreme case — for which there is as yet no long-term solution — of radioactive wastes generated by nuclear power-plants, weapons programmes, and industrial, medical, and research, activities.

In some political jurisdictions, especially in the industrialized countries, controls and waste-processing requirements have been established to reduce pollution and other environmental deterioration, sometimes in ways that add to production costs. Responses to these spiralling complications have ranged from reuse/recycling programmes to industrial waste-limitation processes, emission controls, and even the emergence of an advanced waste-management industry. One additional response is to export wastes, whether because near-by disposal sites are saturated or because it costs less to ship wastes than to stop generating them. Some wastes are moved to regional storage and management facilities (as with low-level radioactive substances) or to relatively open areas within a country, and some wastes from industrialized countries have been exported to relatively poor countries — a practice that has been characterized as a type of economic colonialism.

Urban development provides the leading example of the growing, changing, and increasingly complex, interactions among humans and between humans and their environments; the worldwide urban situation also demonstrates some social and health problems that result from the development process under conditions of population growth. Cities' involvement with one another and with their own and other ecocomplexes* provides clues to the nature and structure of a 'new' Man–environment system — one that has been emerging at an ever-faster rate over the last half-century. Only abstractly and artificially can that system be viewed as a purely Man-made entity — or, rather, complex of entities*; to be realistic we must recognize its intimate, pervasive involvement with Nature at every point and at every level, from the individual household to the entire Biosphere.

HUMAN REPRODUCTION, HEALTH, AND ENVIRONMENT

Child Survival and Community Health

Where records in industrialized countries show rising levels of health, as indicated by increased longevity and falling infant-mortality rates, improvements are most closely associated with gradual advances in three environmental conditions: nutrition, hygiene, and slowing natural increases in population; the last of these involves changes in reproductive behaviours that protect people against hazards in the physical environment. Better nutrition is a function of more knowledgeable use of natural resources in agriculture (embracing also horticulture) and a social environment that is adequate to support production, distribute food, and enable people to buy and use it well. Reproductive behaviour that spaces and limits the numbers of pregnancies is strongly influenced by a social environment that takes account of improved birth outcomes

---

* Dominated in their component parts by different concerns and, to that extent at least, social counterparts of the 'natural' ecocomplexes, *cf.* N. Polunin & E.B. Worthington: 'On the use and misuse of the term "ecosystem" ', *Environmental Conservation*, **17**(3), p. 274, 1990. — Ed.

and infant–child survival by altering the norms for the size and character of the family. Clearly, the three conditions are interactive and mutually reinforcing. Mothers whose health has not been compromised by too many and too frequent pregnancies can produce healthier infants. As people come to realize that fewer children than formerly die before attaining maturity, they are less likely to 'over-reproduce' in order to ensure that the personal goal of family size will be met. When the family and the community require fewer births to meet their respective demographic norms, each experiences a smaller drain on its own resources; at the same time, fewer persons to be nurtured and sustained means a decreased drain on the environment — what is taken from it, what wastes it must absorb, and what alterations it must undergo (leaving aside the matter of social norms of consumption).

This personal/local model of the population issue has been widely adopted in national and global policies and strategies, which have critical implications for improved health and environmental preservation. Policies that seek limited or zero population-growth — that is, a population replacement policy — are recognized as the most basic requirement in realizing the goal of sustainable development. Hypothetically at least, communities have options in limiting wasteful production and overconsumption; but when once a human being is born, he or she is considered to have a rightful demand on the resources needed for life, so that every viable birth constitutes an irrevocable claim on the social and physical environment.

In actuality, of course, social systems are not so easily modified and, for those systems to be preserved, human life may well be sacrificed to economic values, to cultural beliefs, and to class, ethnic, and national or sometimes even religious, interests. Also, policies that seek the continuing growth of economic systems — for whatever motives — implicitly require that populations increase in order to provide more production labourers and more consumers in ever-expanding markets. Such economic growth-policies are usually accompanied by rising consumption expectations, fuelled in part by legitimate desires to satisfy basic needs for decent and secure living-conditions. But people's expectations rise also in response to being exposed to the extravagant visions of life-styles projected by the increasingly global mass-media.

Population Growth and the Global Commons

While the above trends can be measured — and their impact on the shared resources of The Biosphere estimated — at the global level, the actions that make up the trends are taken by multitudes of actors who function mainly in isolation from one another. For example, 'over-fishing' has brought about a global problem of declining stocks of food-fish, but over-fishing happens because countries and entrepreneurs each exercise a 'right' to draw upon ocean resources without regard to what others are doing; forests have been destroyed by people seeking to satisfy their own short-term needs and wants, without considering the consequences for the survival of ecosystems or future generations. On the demand side, countries which pursue policies that seek to influence reproductive behaviour must contend with an apparently-innate sense that any person has a 'right' to procreate, without regard to social or environmental consequences. In many countries, defence of these 'rights', traditionally sanctioned, has taken many forms — ranging from passive resistance to open, passionate political conflict and even acts of homicide.

The deeper conflict, however, is between established ways of dealing with social needs and the ways that must be found to keep already deteriorating environmental life-supports from sliding into catastrophe. If traditional ways continue, the mere extension of current demographic, developmental, economic, urbanizational, and technological, trends will create ever-more-severe dangers to human health, to its supporting natural environment, and to a tolerable social environment. Because these large dangers are generated from relatively small and dispersed sources, they grow by encroachment and marginal change; only rarely do they manifest themselves as visible, attention-commanding disasters.

Although an apocalypse cannot be ruled out, considering worst-case predictions of global warming and its consequences, the more likely manifestations of these increasing dangers are likely to be falling living-standards and more widespread poverty, increasing incidence and actual incidents of famine, ever-larger refugee populations, rising levels of disease, acceleration of community and international violence, and progressive breakdowns of government and communal systems, with an increasing prevalence of a 'politics of hate' as currently exhibited in Bosnia-Herzegovina, Somalia, Rwanda, Zaire, Liberia, Ireland, Korea, and even elsewhere.

Of special health-concern are the effects of climate changes on the mutation of communicable-disease pathogens and the distribution of vector populations. As increasing numbers of people find themselves at the margin of survival, pressures on the natural environment are bound to increase rather than lessen, and social environments will be impacted economically and politically by people's inevitable and ultimately unstoppable movements to cities and countries that offer some hope of enduring. Communities might also be forced to confront painful choices around the issue of the comparative worth of human lives not only as between 'natives' and 'aliens' but also between persons in the productive years of life and those at the extremes of the age-spectrum. Without substantial modification of the trends noted above, the lot of the human species must be expected to move progressively towards the famous characterization of life by the English philosopher, Thomas Hobbes, as 'nasty, brutish, and short'.

Modifying the Trends

In recent ecological dialogues, four major factors have been identified as holding the key to reversing or containing the trends in environmental change, at least as they influence human well-being. The first is to pin Humankind's hopes to the development of technologies that will counter the adverse impacts of profligate growth and production and also provide substitutes for natural resources and conditions — especially resource-substitutes to meet Humankind's needs for food and energy. The second is to rely on the Planet Earth's own mechanisms for 'self-correction' or adaptation. The third concerns the demographics of population growth and urbanization, while a fourth involves the character and level of economic consumption.

The degree to which future development and application of *'clean' and substitute technologies* can provide a way out of the quandary is, of course, unknown. For example, practical mastery over the use of power from nuclear fusion might, sooner or later, reduce the atmospheric assaults and non-replaceable fossil-fuel drains involved in electric energy generation for residential and manufacturing use. Bioengineering might, sooner or later, assist in meeting the food needs of larger human populations, and aquaculture could also contribute. Less feasible in significant numerical terms — and qualitatively much less satisfactory — is extraterrestrial colonization. But each of these contributions remains conjectural, not only because of the uncertainties of technical success, but also of questions of economic feasibility considering the magnitude of the investments required and the reluctance of governments to support scientific research in times of financial stringency.

This approach is also fraught with political and social issues, including the time, expense, dislocations, and other difficulties, of converting large populations from relatively simple technologies and moving them onto a sophisticated technological stage. Even before that problem is reached, however, radical changes in attitudes, policies, and laws, would be necessary to enable the free flow of technological information among countries and organizations — to mobilize the needed investments and distribute them equitably; to establish meaningful agreements on goals and priorities; and, ultimately, to transform the basic motivation of self-interest into shared purposes.* Each such change would entail not only difficulties, but also some infringement of personal freedom, some intrusion into the 'sphere of anarchy' that has been, on one hand, a root factor in bringing on the crisis and, on the other, a source of human creativeness, adaptability, and spiritual growth. One need only consider a world in which human reproduction is technologically controlled by political elites to envision a *qualitative* death of the species.

Imaginative 'ecologists' hypothesize a capacity of the planet and its natural systems to *'self-correct'* — to adapt to changes in such a way that equilibrium or homeostasis is restored. The phenomenon has been observed in the recovery of ecosystems and species after devastating fires and volcanic eruptions; it is assumed that the stratospheric ozone layer will somehow regain its capacity to shield against ultraviolet light, if and when human releases of well-known destructive chemicals are curbed. To the extent that this hypothesis is true, what is not known is whether all natural systems have this capacity of recovery, or how the required restoration-time relates to even minimum biotic needs that are dependent on the damaged systems. Nor do we know the degree to which self-correcting capacities of some natural systems have already been exceeded; for example, although human-based releases of 'greenhouse gases' is small relative to that from volcanic and other natural sources, the anthropogenic increment added in the last half-century (or likely in the next) may be greater than homeostatic processes can correct. We do not know, in short, the extent to which our 'breathing room' has already shrunk. Applied perversely, reliance on self-correction mechanisms implies letting 'Nature take its course' through famine, disease, and/or conflict, to reduce human populations to levels that ecosystems can support — as of course occurs with other species whose numbers exceed natural bearing-capacities.

Population projections that assume moderate success through family planning programmes nevertheless indicate the almost-certain doubling of the planet's 5.8 thousand million human inhabitants by the latter part of the 21st century, as birth-rates continue to fall more slowly than death-rates. These, as well as more and less optimistic projections, recognize that the progenitors of this population-increase are already alive, given the number and the age-

---

* Here we are reminded of our long-time thought — which has been pronounced by an eminent geneticist as quite feasible — of the need of the genetical-manipulation, cloning, or emergence otherwise, of a dominant strain of humans who would always place the welfare of The Biosphere above all selfish considerations and personal ambitions. Meanwhile, despite increasing talk of there being 'only one Earth' and the telling slogan of 'Save Our Biosphere' being carried on the cover of the Indian journal *Environmental Awareness*, it is disappointingly rare to find anybody who sees much beyond his or her own horizon or really cares for much more than family and local interests. — Ed.

profile of the world's population. Also recognized is that most of this increase will occur in the poorer, so-called 'developing countries' — much of it in their cities and under conditions which are highly adverse to health and environmental integrity, as discussed above. Some 30 years of experience with family planning programmes (and other approaches) indicates, on one hand, that progress can be made by assisting many communities to obtain effective access to contraception and that family education can lead to conserving choices; on the other hand, that experience also demonstrates the power of such contrary forces as religious and prejudicial opposition, ignorance, political and ethnic agendas, sexual inequality and stereotypes, and community traditions. Although the progress made in lowering birth-rates is encouraging, the likelihood of solving global problems of health and environment through population limitation alone is low. With respect to the trends of rapid urbanization in developing countries, the history of efforts to retard or reverse them is shorter and bleaker, though most countries are now seeking policies and action programmes which enable their urban residents to cooperate effectively towards improving their living conditions.

But even if the demographic factors could be controlled, the quantity and quality of *consumption* could still lead to disaster. The objective is to ensure that finite resources are not wastefully depleted, and that increasing pollution is avoided. This requires a reduction in unsustainable life-styles in the industrialized countries and methods to improve the impoverished condition in less-developed countries without stressing the environment in the manner and to the extent that the more affluent nations have done. A multi-pronged strategy must modify demand and reduce waste and pollution, even while providing for the basic needs of hundreds of millions of the world's poor.

Some national experiences provide clues to reducing people's energy and other resource-draining demands, and some progress has been made in developing 'clean' technologies and providing 'environmentally-friendly' products. Still greater progress is needed in these aspects of development, in the free transfer of technical know-how within and among countries, and in developing popular acceptance and cooperation on the part of people in all countries. Making that progress faces more than technical difficulties: the greater task is to effect successive changes in people's expectations and social identifications, in their exercise of no more than proper rights, and in their adjustment to the inevitable economic dislocations that altered consumption-patterns entail.

Even though no one major factor can be manipulated to contain or reverse the adverse trends, simultaneous and coherent action on all of them (and on related factors) might have substantial effects. To do so requires a global agenda, operational at local and intermediate levels, which seeks to minimize further environmental insults and correct damages. That agenda would aim at reducing *per caput* and aggregate demands on Nature and its resources, use sound production and waste-management technologies in satisfying remaining demand, and undertake positive programmes of environmental restoration and resource replenishment. In general, that is the doctrine of sustainable development, on which the 1992 United Nations Conference on Environment and Development (UNCED) made noteworthy progress in broadening political acceptance, conceptually integrating concerns about Nature with those about human well-being, and defining lines of action, at least in sectoral terms. In these respects, that Conference set the stage for the enormous and immensely difficult tasks of implementation.

Making Holism Work

The essence of the difficulty is to translate the commitments of UNCED and other global gatherings into consistent, day-by-day activities in countries and communities — also to sustain the enthusiasm and animate the concepts. The roots of the difficulty are found in our human history of *chronic social inertia in acting for other than selfish reasons*.

Only lately have countries begun to work out patterns of internal and international cooperation relevant to inhibiting or reversing the environmental trends that are so menacing to health and welfare. Most such efforts, whether national family-planning programmes or multilateral fishing treaties, entail the resolution of conflicts among competing, established interests. Although selfish economic interests are usually the most visible, social and cultural interests are inevitably also involved; some interests are overt (cultural intrusions are always resented), and others are concealed (policymakers in many societies, for example, strongly oppose enlarging women's role in decisions about their reproduction).

In the fora of national politics and international diplomacy, conflict resolution has been difficult, time-consuming, and frustrating — most often because of the different contexts and situations from which negotiators come, and the imperatives that derive from the priorities 'naturally' accorded to local, class, and national, interests over the larger, 'more abstract' interests of our planet and species. Domestic political concerns are also served by the established international mechanisms that often deal separately with factors which in reality are interlocked in making up the whole of the crisis of environment and human well-being. Thus, for example,

UNCED duly noted the significance of population and other social factors in sustainable development — including habitat and the situation of women. It also requested a change of consumption patterns and the combating of poverty. But its primary action proposals, and all proposals with high political visibility, concerned issues in the environment without mention of population.

The international fragmentation of the problem is mirrored universally at national and local levels, where the specialization of responsibilities and action, generated mainly by Western traditions, infests science, production, the professions, and government. Although specialization has generally been highly efficient in moving Humankind forward in science, technology, and economic production, its uninhibited practice presents a stumbling-block to the solution of current and impending problems. How to harness the power of specialization to the guidance of a holistic sense of the crisis, its structure, and solution, is the crucial task for the immediate future.

The best point of departure in that task may well be to focus it on the highest stakes of all: the integral relationship between the health of the planet and the health of people.

SUPPORTING DOCUMENTS

Agenda 21 (1992). United Nations Conference on Environment and Development. (Rio de Janeiro, Brazil, 3–14 June 1992.) United Nations, New York, NY, USA: 294 pp.

Environment and Health: The European Charter and Commentary (1990). First European Conference on Environment and Health (Frankfurt, Germany, 7–8 December 1989). World Health Organization Regional Publications, European Series Nr 35, Copenhagen, Denmark: x + 154 pp.

Environmental Health in Urban Development (1991). Report of a WHO Expert Committee. (WHO Technical Report Series Nr 807.) World Health Organization, Geneva, Switzerland: 71 pp.

HANLON, JOHN J. (1969). An ecologic view of public health. *American Journal of Public Health*, **59**, pp. 1–11.

Health and the Cities: A Global Overview (1992). Document A44/Technical Discussions/2, World Health Organization, Geneva, Switzerland: 35 pp.

Health, Environment, and Development: Approaches to Drafting Country-level Strategies for Human Well-being Under Agenda 21 (1993). Document Nr WHO/EHE/93.1, World Health Organization, Geneva, Switzerland: 39 pp.

KENNEDY, PAUL (1993). *Preparing for the 21st Century*. Random House, New York, NY, USA: 512 pp.

Oslo Declaration on Environment, Health, and Lifestyle (1991). 30th IULA World Congress, 23–27 June 1991, International Union of Local Authorities: in English, French, and German, each 10 pp.

Pan American Charter on Health and Environment in Sustainable Human Development and Regional Plan of Action (1995). Pan American Conference on Health and Environment in Sustainable Human Development; Meeting of Ministers of Health, Environment, and the Economy (Washington, DC, USA, 1–3 October 1995): 4 pp. and iv + 11 pp., respectively.

SCHAEFER, MORRIS (1991). *Combating Environmental Pollution: National Capabilities for Health Protection*. World Health Organization, Document WHO/PEP/91.14, Geneva, Switzerland: vi + 38 pp.

Senanayake, P. (1995). Reflections on the UN's Conference on Environment and Development. *Environmental Conservation*, **22**(1), pp. 4–5.

The Sundsvall Statement: Supportive Environments for Health (1991). (Declaration of the 9–15 June 1991 Conference Jointly Organized by the World Health Organization and the Nordic Countries, in Association with the United Nations Environment Programme.) Now published as: *Supportive Environment for Health: Major Policy and Research Issues Involved in Creating Health Promoting Environments*, by Kathyryn Dean & Trevor Hancock, WHO Regional Office for Europe, Copenhagen, Denmark: 42 pp.

WHO Global Strategy for Health and Environment (1993). Document WHO/EHE/93.2, World Health Organization, Geneva, Switzerland: 28 pp. + 12 pp. of (3) annexes.

World Commission on Environment and Development (1987). *Our Common Future*. Oxford University Press, New York, NY, USA: xv + 400 pp., tables.

World Health Organization (1992). *Our Planet, Our Health: Report of the WHO Commission on Health and Environment*. WHO, Geneva, Switzerland: xxxii + 282 pp., illustr.

[Added in proof from page 143]

* The World Health Organization's main headquarters being within easy walking distance of our abode and office, one of our most memorable experiences was in 1980 when we were visited by an Australian friend who was Chairman of WHO's Global Commission for the Certification of Smallpox Eradication, who brought us the news that they had just confirmed that the dread disease no longer existed apart from specimens of the causal organism preserved in a very few known and controlled laboratories. So our celebrating drink may well have been the first to mark such a major advance of humankind towards improving its environment. — Ed.

8

World Population
and Nutritional Well-being

by

VALERIA MENZA, *Nutrition Officer*

&

JOHN R. LUPIEN, *Director*
Food Policy and Nutrition Division,
Food and Agriculture Organization of the United Nations,
Via delle Terme di Caracalla,
00100 Rome, Italy

INTRODUCTION

Providing for the nutritional well-being of a rapidly growing world population is one of the greatest challenges facing the global human community. The deep concern that is being expressed world-wide about our ability to meet this challenge is understandable, given the projected increase in the world population from 5.75 to 8.47 thousand millions between AD 1995 and 2025. Even greater cause for concern is the fact that almost all of the increment is expected to occur in the less-developed regions, where 20% of the population — 841 million people* — is estimated to be chronically under-nourished[†], having insufficient access to enough food to meet their basic daily needs for good health and nutritional well-being. The 80 to 85 million people who are expected to be added annually to the numbers in the 'developing world' will need health, nutrition, and population, services, as well as educational and employment oppor-tunities, if fertility[§] rates are to be lowered and their nutritional well-being duly ensured. The increases in world population and rapid rates of urbanization are critical issues in terms of food availability, access to food, and nutritional well-being; more people will require more

---

* The Sixth World Food Survey (FAO, 1996); (*cf.* also FAO/WHO, 1992*a*, 1992*b*, 1992*c*; FAO/ESS, 1993).

† Defined as those people whose estimated daily energy intake over a year falls below that required to maintain body-weight and support light activity.

§ Fertility here and hereafter being measured by the number of children born to a woman. — Ed.

157

food, more goods, more services, and more employment opportunities.

Much of the debate on our ability to provide for, and the Earth's capacity to support, this number of human beings, has focused on whether or not world food production will be able to increase by as much as will be necessary to maintain sufficient food supplies for the growing population. However, increasing global agricultural production to meet the needs of such a growing world population, while a prerequisite for eliminating hunger and malnutrition, is not sufficient by itself to solve these problems. This is borne out by the fact that hunger persists today even though world agriculture produces enough food to meet everyone's needs. If the global food supplies currently available were distributed according to individual requirements, they would be sufficient to provide well over what is needed to meet the energy needs of the world population; yet one out of five persons in the 'developing world' continues to be chronically undernourished. While ever-increasing food supplies will be needed to meet the demand created by a growing population, the problems of hunger and malnutrition cannot be addressed without countering the underlying impediments — mainly poverty and underdevelopment — to adequate access to food by all living individuals.

The ability of households to acquire sufficient supplies of a variety of good-quality and safe foods to ensure their nutritional well-being depends on both the availability of the necessary food and the means to acquire it. For a number of developing countries — particularly the low-income food-deficient ones — access to food for the majority of the rural population depends largely on domestic local agricultural activities, which provide not only food, but also employment and income. Particularly for the landless rural population, the overriding nutritional problems are more closely associated with a shortage of jobs than with a shortage of food. Further development of the agricultural sector is necessary not only for increasing and improving the food supply, but also for increasing employment opportunities and, therefore, increasing incomes and the purchasing power of the nutritionally deprived, poor, and disadvantaged, groups of the population.

Poverty, social inequality, and lack of education, are primary causes of hunger and malnutrition. Malnutrition primarily affects poor and disadvantaged households, and poor health related to malnutrition compounds the situation by further reducing their already meagre resources and earning capacities, thus increasing their social and economic problems. This, in turn, contributes to further declines in future human, economic, and social, development. Prospects for improving the food and nutrition situation in developing countries is likely to depend on the prospects of those countries for raising incomes, reducing poverty, and improving the overall social and econo-mic conditions. In many societies, the subordinate position of women may also be an important cause of malnutrition; children and women are among the first victims of poverty and malnutrition in situations where prevailing beliefs consolidate their position of inferiority.

Disquietingly, the number of people in developing countries liv-ing in absolute poverty is increasing; some estimates indicate that the number of poor rose from 1.2 to 1.3 thousand millions between AD 1990 and 1991 (UNDP, 1993). As the poor people in most societies do not have adequate access to the basic requirements for nutritional well-being — nutritionally adequate and safe food, safe water and sanitation, health services and education — such an increase gives cause for grave concern for the health and welfare of these people. The consequences of malnutrition for human well-being and for socio-economic development are profound and far-reaching; poor nutrition and health result in a reduction of the overall quality of life and in the levels of development of human potential.

EXTENT OF HUNGER AND MALNUTRITION

Hunger and malnutrition persist as one of the most devastating prob-lems facing the world's poor, and continue to exist in some form in almost every country. One out of every five persons in the 'developing world' is chronically undernourished. Approximately 190 million children under five years of age suffer from acute or chronic protein-energy malnutrition; during seasonal food-shortages, and in times of famine and social unrest, this average number increases. According to some estimates, malnutrition is a factor in one-third of the nearly 13 million children under the age of five who die every year from diseases and infections (UNICEF, 1992, 1993, 1994). In addition, malnutrition in the form of deficiencies of iron, iodine, and vitamin A, continues to cause severe illness or death of millions world-wide. Lack of iron is estimated to affect over 2,000 million people, insufficient vitamin A affects 40 millions, and over 1,000 million people are at risk of iodine deficiency. In many developing countries, 50% of these population-groups are affected by such deficiencies. Various other micronutrient deficiencies, caused by a lack of zinc, selenium, or other trace-elements, affect large numbers of people in some parts of the world. Outbreaks of classical deficiency-diseases — beriberi, pellagra, and scurvy — still occur in refugee camps and other deprived populations, and rickets affects significant numbers of children.

Undernutrition and micronutrient deficiencies may result in a range of conditions that adversely affect the health and well-being of individuals. In severe cases they can be life-threatening. Iodine deficiency, found world-wide, affects particularly populations located in mountainous or flood-prone areas where soils are deficient in iodine. Iodine deficiency not only leads to goitre but also impairs physical and mental development. It is the most common preventable cause of mental retardation. In severe cases, this deficiency leads to deaf-mutism, cretinism, and other serious defects. Over 200 million people have goitre — of whom 6 million are cretins — while 26 millions suffer from mental defects caused by iodine deficiencies. Lack of this micronutrient can impair children's resistance to infection, leading to increased mortality. It can also impair reproductive functions, leading to increased rates of abortion, stillbirth, and congenital anomaly.

Iron deficiency affects people living in practically all countries. The people most affected by iron deficiency are women and children of pre-school age — often more than 50% of those are anaemic, although older children and men may also be affected by lack of iron. Anaemia in infants and children can retard physical growth and cognitive development and lowers resistance to infections. In adults, iron deficiency causes fatigue, lowers work-capacity, and impairs reproductive functions; as many as 20% of maternal deaths are caused principally by iron deficiency.

Vitamin A deficiency (VAD) occurs especially in areas where fruit and vegetable consumption, and sometimes fat intake, are low. It is the most common cause of preventable childhood blindness; a quarter- to a half-million children become totally or partially blind every year from vitamin A deficiency, and two-thirds of these children die within months of going blind. VAD has a wide range of effects, reducing the effectiveness of the immune system and retarding growth and development. Even moderate levels of deficiency can lead to stunted growth, increased severity of infection, and higher death-rates.

Changes in diet and life-style associated with urbanization, higher incomes, and longevity, have led to the emergence of diet-related non-communicable diseases — such as obesity, cardio-vascular diseases, diabetes, and some forms of cancer — as major problems. Hundreds of millions of people also suffer from diseases caused by contaminated food and water (FAO/WHO, 1992*b*).

TRENDS and ACHIEVEMENTS

The region with the highest proportion of its population undernourished is sub-Saharan Africa (41%). There, the number of people affected has more than doubled since the 1970s (from 103 million people to 215 millions between 1969–71 and 1990–92), although the proportion has increased only slightly (from 36%), primarily as a result of the region's annual population growth-rate. In Latin America and the Caribbean, where the proportion is low (14%) and continues to decrease, the *number* of undernourished has also increased as a result of population growth. Currently, the greatest number of chronically undernourished live in Asia (512 million people). While the percentage of underweight children under the age of five has been declining over the last 15 years on all continents, the absolute numbers of underweight children have remained fairly stable as a result of population increases. In 1990–92, two out of five children in the developing world were stunted (low height for age), one out of three was underweight (low weight for age), and one out of ten was wasted (low weight for height). In absolute numbers, there were 215 million stunted children, 179 millions underweight, and nearly 90 million wasted children, in 1990. In Africa the number of underweight children increased from 20 to 27 millions between 1975 and 1990.

Encouragingly, much progress has been made in reducing the prevalence of nutritional problems, and many countries have been remarkably successful in addressing issues of hunger and malnutrition. Globally, the number of people suffering from chronic malnutrition has declined consistently over the last twenty-three years of calculation (from 918 million people to 841 millions between 1969–71 and 1990–92), as has the proportion of undernourished people (from 35% to 20%), even though the world population has increased. Furthermore, current projections indicate that this trend will continue. In East and Southeastern Asia, striking improvements have occurred in the last 20 years, with the proportion of the population affected by undernutrition declining from 41 to 16%. Life expectancy in most developing countries is improving rapidly, mainly as a result of reduced early deaths from infectious diseases, though mortality rates among children are also declining. Infant mortality rates, and to a lesser degree maternal mortality rates, have also declined in many countries.

In Thailand during the last decade, the prevalence of protein-energy malnutrition (PEM) among pre-school children declined dramatically from 50.8% to 17.1%, with almost total elimination of moderate and severe forms. This success is largely attributable to the country's main five-years social, health, and food and nutrition, plans, and especially to the Poverty Alleviation Plan (PAP) concentrating on raising living standards to a subsistence level and providing minimum basic services in high-poverty areas.

In Indonesia, food availability increased from 2,106 to 2,697 kcal per person between 1971–73 and 1990–92, and the prevalence of malnutrition is decreasing steadily. Rapid and equitable economic growth and increased food availability are responsible for these improvements; agriculture grew by over 4% per annum in the 1970s and by over 3% per annum in the 1980s, and the country is now self-sufficient in rice due to Green Revolution technology, production incentives, rural credit, and sound marketing practices (Tontisirin, 1992*a*, 1992*b*).

Chile has achieved remarkable advances in improving the health and nutritional status of infants and pre-school children over the last three decades, despite the persistence of poverty, under-development, and severe economic and political crises. Both infant and child mortality rates have declined from one of the highest in the region to one of the lowest; the prevalence of child malnutrition has declined from 37% to 8.5%. These achievements have been made through well-targeted policies and programmes in health, sanitation, education, and food production. Some researchers maintain that 20% of the reduction in the infant mortality rates can be ascribed to the decrease in birth-rates since family planning services were initiated (Monckeberg, 1992). Agricultural policies initiated in the mid-1970s have resulted in a complete turn-around in this sector, and food production has increased rapidly. This has led not only to a sharp decline in costly food imports but also to a substantial increase in rural employment and income.

The elimination of famine in India is a major achievement of policies of food security over the last two decades. The overall growth in food availability resulting from Green Revolution technologies and a substantial reduction in poverty are responsible for this. China has made remarkable advances in agricultural production and overall economic growth which have contributed significantly to improving the nutritional status and living standards of its people. Faced with a growing population (currently 1.1 thousand million people, or almost one-fifth of the total world population) and relatively little arable land, China has formulated a number of policies aimed at increasing food production and supply and improving distribution to all levels of the population. Particularly successful have been those policies focusing on the development of the rural economy. Great emphasis has been placed on the primary health-care network, through which much progress has been made in nutrition education for the prevention and control of malnutrition among rural children and the prevention of non-communicable diseases among the urban population (Ge *et al.*, 1991, 1992).

In Brazil, national averages of the prevalence of underweight children fell from 18.4% to 7.1% between 1975 and 1989. Substantial improvements have been made in the nutritional status of pre-school children in Zimbabwe, where infant mortality rates have declined sharply. Botswana, despite current and persistent drought, has eliminated deaths from famine and starvation. These accom-

plishments reflect improvements in food availability and consumption, health care and sanitation, and increasing income — all of
which are fundamental aspects of development.

FOOD SUPPLIES AND AGRICULTURAL GROWTH

The world as a whole has made significant progress in increasing
food supplies. World *per caput* supplies of food for human consumption are today 18% above what they were 30 years ago. In the
1980s, average *per caput* food supplies continued to increase in
developing countries, although at a slower rate than in the 1970s.
This generally held true across the board for low-income countries,
with the exception of the People's Republic of China, where food
availability increased considerably in the 1980s. However, this
progress has been uneven, by-passing many countries and population groups. Little progress was made, on average, by the group of
the least-developed countries during the last decade. The world
agricultural growth-rate has been consistently slowing down over
the last three decades (from 3.0% per annum in the 1960s to 2.0% in
the period 1980–92), and this decline appears to be even more pronounced in the eight-years' period between 1984 and 1992 (1.7%).
The decline in the rate of increase in world cereal production — the
main staple food-source for most populations — has been even
sharper (from 3.6% per annum in the 1960s to 1.6% in AD 1980–92)
(World Agriculture, 1995).

The situation is particularly serious in sub-Saharan Africa, where
severe food-supply difficulties are faced continually. Recurrent
drought and, in some cases, ongoing civil unrest, regularly depresses
food availability from often already unacceptably low levels. The
continued civil conflict in a number of countries has not only curtailed domestic food production, but has also led to internal displacement of refugees and has hampered efforts to provide relief to
those persons who are most affected. Drought also affects the
livestock sector, upon which the livelihood of a large part of the
subregion depends. Overall, approximately 40 million people in the
region are dependent on continued emergency assistance for their
survival.

Despite the decline in food availability in several countries
during the 1980s, by the end of the decade roughly 60% of the
world's population was living in countries that had more than 2,600
calories (kcal) available per person per day. While malnutrition still
exists in countries with adequate aggregate food-supplies, an
inadequate food-supply clearly indicates an even higher prevalence
of nutritional problems. During the period 1988–90, there were
about 11 countries, with a total population of 123 million people,

with *per caput* daily energy supplies (DES) of less than 2,000 kcals per person. It is not possible for a population to meet its energy needs from such a limited supply; hunger and malnutrition are inevitable among many of the people living in these countries where the food situation is so critical.

It is foreseen that the growth-rate of gross world agricultural output may slow down further to 1.8% per annum in AD 1988/90– 2010. Even so, it is expected to continue to outpace the 1.5% annual growth-rate of the world population, although the growth-rate of *per caput* production may also be lower than in the past. However, the impact of the slow-down in agricultural production on access to food by all population groups, and in particular by those already inadequately nourished, cannot be evaluated at the global level. Aggregate estimates of food production and availability at either global, regional, or country, level cannot reflect household or individual food consumption. At the household level this depends on the ability of households to produce or procure food, and is a function of income levels and distribution, food availability and wastage, as well of prices and consumer choices. At the individual level, social status within the household, preferences, and care and feeding practices, are also important determinants of dietary intakes.

One interpretation of the significance of the projected further slow-down in world agricultural growth for food security — and, consequently, for nutritional status — uses the criterion of a country's share of agricultural labour-force in its total labour-force, emphasizing the demand or income aspect because it measures the share of the population that is dependent mainly on agriculture for employment and income. The majority (62 out of 94) of the 'developing countries' for which production estimates are available, can be classified as having a high dependency on agriculture, with over one-third of their economically active population engaged in agriculture*.

All except four of those 64 countries currently have food supplies under 2,600 calories *per caput* and the majority of them are closer to 2,000 calories — levels insufficient for good health and nutritional well-being. Moreover, the majority of those countries have comparatively low net food imports *per caput* and some depend heavily on agricultural exports for their balance of payments, and thus on their capacity to import food. The food security of these countries depends heavily on their agricultural performance both from the demand side — depending on the income of the bulk of their population — and from the supply side, involving food that is self-produced or imported using agricultural export earnings. Projections

---

* For a complete explanation of this analysis, *see* World Agriculture (1995).

foresee no slow-down, but perhaps a modest acceleration, in the growth of *per caput* production in this group (excluding China), indicating that there may not be a deterioration in the growth-rate of production in relation to population growth but that agricultural growth will not be sufficient to result in significant improvements in the food security and nutritional well-being of those populations.

NUTRITIONAL STATUS AND FERTILITY

Concern is often expressed that, along with improved nutritional and health status, could come the potential of higher fertility rates and increased family size, as reproductive periods are lengthened and child survival increases. However, experience shows that as families improve their health and nutritional status and child survival increases, and as the family unit becomes more economically stable, fertility rates decrease.* As parents become able to count on the survival of their children, and become less dependent on them as sources of labour, they have fewer children. Better nutritional status and lower infant mortality tend to slow down population growth, which, in turn, improves child and maternal health, and ultimately improves the overall health and welfare of the entire family. Policymakers concerned with the impact of population on development, must recognize that fertility rates and nutritional status are inextricably linked.

Improving nutritional status is a key aspect of lowering fertility. Research over the past decade presents mounting evidence that poorly-nourished women give birth to underweight babies, who often have developmental problems. In the worst cases, these children have a poor chance of survival (Shetty & James, 1994). The age at which a woman first becomes pregnant, her spacing between pregnancies, and the number of pregnancies she has, all affect her health and nutritional status and those of her children. The biological demands of pregnancy and lactation, compounded by the physical demands of their work, can create nutritional problems for women which can also affect subsequent pregnancies. Breast-feeding has a fundamental role in improving nutritional well-being, as it provides infants and young children with the ideal nutrition and has a positive influence on birth-spacing.

Iron-deficiency anaemia in infants and children is associated with retarded physical growth and development of cognitive abilities and low resistance to infections. In women it can cause grave impair-

---

* Similar points are made repeatedly in the present volume — up to the last chapter which practically ends with a figure (on page 292) indicating the ever-increasing 'Percentage of Women Wanting No More Children'. — Ed.

ment of reproductive functions. Maternal anaemia predisposes women to haemorrhaging and contracting infections prior to, during, or after, childbirth, and as many as 20% of maternal deaths are caused principally by iron deficiency. Maternal anaemia also leads to intra-uterine growth retardation, low birth-weight, and increased rates of perinatal mortality.

The health risk to both mother and child is particularly serious among adolescent mothers; if a pregnancy occurs before the mother's requirements for growth have been adequately met, additional demands are created on her body. The competing nutritional needs of pregnancy and growth in the adolescent mother will affect the growth of the foetus and hence the growth of the child. In both developed and developing countries, pregnancy among adolescents has serious personal consequences for both mother and child, as well as significant costs to society.

STATUS OF WOMEN

Raising the status of women is essential for reducing population growth-rates and improving nutritional status[†]. Maternal education and literacy have a significant impact on children's survival, health, and nutritional well-being. Providing women with increased educational opportunities can have a significant impact on reducing the number of children a woman will have, and can delay the age of first pregnancy for most women. Furthermore, educated women are more likely to work in the paid labour force, and those who work outside the home tend to have fewer children. Generally, a woman's control over her income is also associated with an improvement in the nutritional well-being of her family.

An essential element of good health and nutrition, especially for children, is the level of care provided in the household. In general, care engenders the best use of household food resources and the effective use of all resources to protect children from infection, attend to them during illness, and assist others in the household who may not be able to care for themselves.

In most households, women are the principal providers of care, and their ability to do so effectively depends largely on their personal health, education, time, and energy, as well as on their control over household resources. Due consideration of women's role, knowledge, motivation, time, and control over resources, is fundamental to improving the family's nutrition. In many countries, it is the hard-pressed and time-consuming nature of women's work, both inside and outside the home, that determines the types of foods

---

† *See* especially Chapter 10 of this book, on 'Women and the Family Planning Imperative'. — Ed.

eaten, the ways in which they are prepared, and the response and care given to the other needs of their children.

Improved maternal education and literacy can influence positively the skills and knowledge needed for successful child-care practices and for decisions on expenditures of time and resources. The sharing of responsibilities for care, and the burdens of work in general among household members, needs to be encouraged; efforts should also be made to educate men about their family's nutritional needs, and men should be encouraged to play a larger role than is customary in activities to improve nutrition. Within the family, food should be allocated in such a way that women and girls are not disadvantaged.

Close examination of the role and responsibilities of women in providing and sustaining family nutritional levels, is required to assist women in fulfilling their roles more efficiently and equitably. Access to maternal nutrition and health-care needs to be improved to lessen the risks attendant on reproduction, the burdens of child-care need to be reduced, women's role in food production and farming needs to be recognized, and overall, their position in society needs to be elevated. Raising the status of women and improving their access to basic services such as education, health care, and family planning, with due access to credit,* technology, and fair wages, can lead to increased personal incomes, accelerated economic development, and reduction in family size.

NUTRITION AND URBAN GROWTH

Over the preceding 40 years, the world's urban population, through both migration and natural population increase, has grown more than threefold, and is expected to continue to increase (UNFPA, 1993). This massive shift of population from rural to urban areas brings with it substantial implications for nutrition, access to food, education, and preventive health-care. At the same time, the number of poor people in urban areas is increasing in many countries; they are particularly vulnerable to economic factors affecting commercial food-markets, as they spend a high proportion of their available cash on food and are dependent on wage-labour for income.

In addition, urban households may lack the social support-network that is available to rural households — particularly to female-headed households and the homeless. Rapid urban growth places tremendous demands on government services in most developing

---

* The modern microcredit movement is to be commended as a worthy attempt at relieving the financial situation of, ultimately, millions of the world's poorer house-wives, especially. — Ed.

countries, particularly due to a rapid increase in the number of poor urban households concentrated in crowded slums and squatter settlements. Many municipal governments lack the revenues and human resources needed for planning and administering such services for their growing populations.

Infrastructures such as water-supply, transport, and waste-disposal, take up a substantial share of the municipal budget in many countries. Although urban areas are generally more privileged in terms of resources, services, and facilities, than are rural areas, for large sections of the urban poor the basic needs of housing, water, sanitation, solid-waste removal, education, transportation, and marketing facilities, are not met. Rapid urban development also places stress on infrastructure and services aimed at protecting food quality, in addition to food production.

For the urban poor, lack of adequate water and sanitation facilities remains a major problem which has severe consequences on environmental hygiene and public health. The prevalence of communicable, gastro-intestinal, food-borne and other infectious diseases, is usually high under such conditions. Interactions between infectious diseases and nutrition are well-known, and environments which increase health-risk exposure also increase the risk of under-nutrition and malnutrition.

For the majority of city dwellers, urbanization has led to improved nutritional status resulting from the increased availability and variety of foods and improved access to health and other social services. Urban diets are more varied than those in rural areas, although total energy intakes are not necessarily higher, and not all urban populations have benefited equally from an increased availability of goods and services.

Procuring adequate amounts of good-quality and safe food at affordable prices is the main problem that poor urban families have to face every day. Difficult access to food prevails in the suburban areas of large cities where migrants tend to settle. Many of these areas are illegal squatter settlements which spring up so rapidly that urban officials are unable to control their growth and provide adequate municipal services. Other problem areas are legalized parts of the cities that become so overcrowded that provision of adequate services is also very difficult. In both cases, poor nutrition is a result not only of low income but also of poor or limited food availability and inadequate sanitation.

Most urban households rely almost entirely on purchases to obtain food, and much of this food may be commercially prepared and sold on the street or in markets. This sometimes applies even to infants' foods. Food markets are often located far from the squatter areas where the poorest live, forcing many to pay higher prices for poorer-quality food at small local outlets. Some urban households have small plots of land available to them where vegetables, fruit, and small animals, can be raised. Encouraging urban/suburban gardening in such 'allotments' is an important function of local government, and can improve the food supplies and nutritional status of the poor.

A large number of schoolchildren and working adults in cities and towns consume meals away from home. The provision of nutritious, ready-to-eat food — especially from street vendors — at affordable prices is an important aspect of the urban diet, as well as a positive factor in the local economy. However, the uncontrolled preparation of food for sale by itinerate vendors can spread food-borne disease, and therefore control of the quality and safety of these foods is required.

A major implication of the growth of cities is that a diminishing number of rural producers must meet the ever-growing demand for agricultural products. If the additional urban population is to be fed from local production, such food production will have to increase accordingly. Currently, the food needs of urban populations in many countries are being met instead by an increasing dependence on imports, often due to the marketing of imported foods at subsidized prices which undercut the production and sale of local products.

CONCLUSION

The agricultural sector is a major source of income and livelihood, as well as the main source of food, for many of the world's poor. As such, it presents the greatest opportunity for socio-economic development and consequently offers the greatest potential for achieving sustained improvements in the nutritional status of the rural poor. Often the most pressing need is for employment creation both on and off the farm, through activities related to agriculture. Agricultural policies can affect nutrition positively through improved food production, availability, processing, and marketing, as well as through increased employment opportunities; they can also have an impact on people's time, labour, and energy utilization, environmental and living conditions, and the nutrient content of the foods produced.

Enormous efforts will be needed in all sectors to provide for and protect the welfare and human dignity of the 10 thousand million people projected for the year 2050. To achieve this, human welfare, including nutritional well-being, must be placed at the centre of all population, social, and economic development, policies. The basic right of every human being to adequate food, as well as to an adequate overall standard of living, has long been recognized, beginning with the 1948 Universal Declaration of Human Rights and most recently reconfirmed at the 1992 International Conference on Nutrition (ICN)*. Impressive improvements have been made over the past few decades in food availability, health, education, and social services, throughout the world — all of which have contributed greatly to improving the nutritional status and raising the standards of living among many populations.

Undernutrition, child mortality, and high fertility, are all apt to result from poverty. Without social and economic programmes to alleviate poverty, human societies will continue to be caught in this vicious cycle of undernutrition, high fertility, and underdevelopment. Successful policies to address population issues must include the promotion of more equitable economic development and the provision of improved access to education, health, and family planning, services. Clearly — and especially considering the increasing demands that a growing world population will make — intensified efforts are urgently needed for more coherent economic, social, and food and agricultural, development, if hunger, malnutrition, and rapid population increases, with their consequent human suffering, are to be reduced and ultimately eliminated.

REFERENCES[†]

FAO (1993). *The State of Food and Agriculture* [SOFA]. (FAO Agriculture Series, Nr 26.) FAO, Rome, Italy: xxii + 306 pp., illustr. [Software diskette.]

FAO (1993a). *Annual Series of Total Population Estimates: 1960–1993*. Rome, Italy.

FAO (1993b). *World Food Supplies and the Prevalence of Chronic Undernutrition in Developing Regions as Assessed in 1992*, Rome, Italy.

FAO (1995a). *World Agriculture: Towards 2010*. Rome, Italy.

FAO (1995b). *Sixth World Food Survey*. Rome, Italy.

FAO (1996). *The Sixth World Food Survey*.

FAO/ESS (1993). *World Food Supplies and the Prevalence of Chronic Undernutrition in Developing Regions as Assessed in 1992*. FAO (ESS), Rome, Italy: 26 pp.

FAO/WHO (1992a). *World Declaration and Plan of Action for Nutrition: International Conference on Nutrition*. FAO/WHO, Rome, Italy: 50 pp. [English, French, Spanish, Chinese, Arabic, Russian].

---

* The ICN, held in Rome in December 1992 (*see* FAO/WHO, 1992a, 1992b), unanimously adopted a World Declaration and Plan of Action for Nutrition, pledging to renew efforts at all levels to reduce global hunger and malnutrition, and to improve the nutritional well-being of all populations. The ICN also adopted nine areas of action in efforts to improve nutrition: incorporating nutrition objectives into development policies and programmes; improving household food security; protecting consumers through improved food quality and safety; preventing and managing infectious diseases; caring for the socio-economically deprived and nutritionally vulnerable; promoting breast-feeding; preventing specific micronutrient deficiencies; promoting appropriate diets and a healthy life-style; and assessing, analysing, and monitoring, nutrition situations.

[†] Can be completed through FAO when necessary. — Ed.

FAO/WHO (1992b). *Nutrition and Development — A Global Assessment: International Conference on Nutrition.* FAO/WHO, Rome, Italy: xxii + 121 pp., illustr. [English, French, Spanish, Chinese, Arabic, Russian].

FAO/WHO (1992c). *Major Issues for Nutrition Strategies.* International Conference on Nutrition, Rome, Italy.

GE, K., CHEN, C. & SHIN, T. (1991). Food consumption and nutritional status in China, Part 1. *Food, Nutrition and Agriculture* (FAO, Rome, Italy), **1**(2–3), pp. 54–61.

GE, K., CHEN, C., SHIN T., & ZHANG, S. (1992). Food consumption and nutritional status in China, Part 2. *Food, Nutrition and Agriculture* (FAO, Rome, Italy), **2**(4), pp. 10–7.

IFAD (1992). *The State of World Rural Poverty: An Inquiry into the Causes and Consequences.* University Press, New York.

MONCKEBERG, F. (1992). Integrating national food, nutrition and health policy: the Chilean experience. Chapter 5, pp. 50–73 in *Frontiers of Nutrition and Food Security.* International Rice Research Institute, Manila, Philippines: xii + 171 pp., illustr.

SHETTY, P.S. & JAMES, W.P.T. (1994). Body Mass Index — a measure of chronic energy deficiency in adults. Chapter 6, pp. 35–42, in *Health and BMI.* FAO, Rome, Italy: xi + 57 pp., illustr.

TONTISIRIN, K. (1992a). Rural development policy and plans in Thailand. Pp. 49–55 in *Integrating Food and Nutrition into Development — Thailand's Experiences and Future Visions.* Mahidol University, Bangkok 10700, Thailand: vi + 233 pp.

TONTISIRIN, K. (1992b). Thailand's food and nutrition policies and plans. Pp. 37–46 in *Integrating Food and Nutrition into Development — Thailand's Experiences and Future Visions.* Mahidol University, Bangkok 10700, Thailand: vi + 233 pp.

UNDP (1993). *Human Development Report.* Oxford University Press, New York, NY, USA: ix + 216 pp., illustr.

UNFPA (1993). *The State of the World Population 1992.* UNFPA (United Nations Population Fund, New York, NY, USA). Nuffield Press, Oxford, England, UK: iii + 46 pp.

UNICEF (1992). *The State of the World's Children.* Oxford University Press, New York, NY, USA: 100 pp., illustr.

UNICEF (1993). *The State of the World's Children.* Oxford University Press, New York, NY, USA: [not available for checking].

UNICEF (1994). *The State of the World's Children.* Oxford University Press, New York, NY, USA: [not available for checking].

WORLD AGRICULTURE (1995). *World Agriculture: Towards 2010, an FAO Studies.* FAO and John Wiley & Sons, Chichester, England, UK: [not available for checking].

WORLD BANK (1993). *Investing in Health (World Development Report).* Oxford University Press, Oxford, UK: [not available for checking].

9

Global Migration:
A Thousand Years' Perspective

by

JOHAN GALTUNG, Dr (h.c. mult.)
Professor of Peace Studies, Universities of Granada,
Ritsumeikan, Tromsø, and Witten/Herdecke,
51 Bois Chatton, F-01210 Versonnex (Ain), France

INTRODUCTION

As its title indicates, this Chapter deals mainly with global migr-ation. By this I mean that is is concerned mainly with massive, collective migration of people, namely real *Völkerwanderungen**
— not the mere trickle of legal and/or illegal individual immi-grants, usually in search of paid work, such as was implied in an April 1992 newspaper report that 2,291 illegal migrants — the February and March 'catch' — that had been expelled from Tokyo, Osaka, and Nagoya. These are still very small numbers, although their composition is interesting: the largest group was Iranian, then came South Koreans, then Malaysians, and then Thais — all Asians (*cf.* Figs 2 & 3).

The decision to migrate is the outcome of push-and-pull forces — push away from the point of departure which is where the people concerned actually are, and pull towards the point of arrival which may be any point other than 'here', or at least 'not now'. The first factor would indicate a deficit in livelihood up to the point of suffering; the second factor implies a surplus in knowledge, imagin-ation, and networks. In other words, these are factors that do not come together so often, which partly explains why there is not far more migration. But the basic explanation of this last point is that people usually do not want to move: they prefer to 'stay home'.

---

* I am thinking of the order of magnitude of 10^9 (one thousand millions = 1 Amer-ican billion) by the first quarter or third of the next century, or at latest by AD 2050. This is the estimated number already living in misery (certainly relative, but too often absolute) according to the UNDP *Human Development* reports. This is also the prognosis of the Henri Dumont Institute of the Red Cross in Geneva — *see Frankfurter Rundschaer*, 20 November 1987, predicting an earlier arrival at such a situation.

173

For massive migration to take place, the suffering has to be massive and the attractiveness also has to be considerable (at least relatively). There is no difficulty in identifying places conforming to these conditions with the world in its present state, and with international systems in particular supporting ever-decreasing liveli-hood in some parts and yet ever-increasing consumption in others. But this has been the case for millennia — indeed at least since Humankind mastered the conditions of sedentary civilizations, with accumulation of wealth in the cities, though that was often at the expense of the non-city (*see* References). War is a frequent indirect cause of migration.*

COMPARATIVE CONSCIOUSNESS NEW AND UNADDRESSED

This has followed rather naturally from increasing awareness of the extreme differences in livelihood — brought about by the mass media, with radio and TV also catering to the illiterate — and by tra-vel, as well as exposure to the accounts of those who have travelled, with their photos and, nowadays, videos. Very few places in the world are sufficiently remote not to be thus exposed. This means that livelihood deficits are not only measured in absolute terms, as insults to the basic needs for survival, well-being, freedom, and identity; but they are also felt in comparative terms, as indicating that some have far less — in other words, there is gross *inequality*. Hence have been born the *basic-needs project* to help those in real misery and the more socialistic *equalization project* of redistribution from the most to the least privileged.

What this means is that the alternative to massive migration is massive basic-needs satisfaction, starting with the most needy, and massive equalization, starting with the least privileged — the have-nots — preferably in a sustainable way. Normally this is called 'de-velopment', at least in the less-developed countries. But there are no such world-scale development projects under way, nor any more modest basic-needs projects, let alone more ambitious equalization projects. Only some tiny efforts exist here and there (*see* References).

One reason for this continuing state of affairs is an underlying further non-project, namely a concerted thrust for the removal of the basic structural impediments. The major impediment is the pre-dominant structure of world, regional, and national, economies, engendered as they are by the world market system. Capitalism pro-claims the freedom of all to use property to make more property; then takes it away by penetrating into the periphery, buying up property everywhere, and destroying the environment. Socialism, on

---

* Addition of this last brief sentence, sanctioned by Professor Galtung, enables us to record our recent persuasion of a retired friend to write a book on what we see as the *imperative of outlawing all arms and ammunition manufacture* except for rigidly-supervised police purposes if there is ever to be any real hope of lasting peace in our overpopulated and widely unstable world. — Ed.

the other hand, tries to prevent this from happening by building barriers protecting the periphery, then takes it away by making it state property. Both behave in the manner of hypocrites and swindlers. The condition is deteriorating all the time. Of the three major efforts to counteract these massive structures of inequity one, the indigenous project, is dying out; another, communism, has collapsed; and the third, 'green' economics, has not yet come off the ground in a sufficiently massive way for anything like global effectiveness. Should it take off, my prediction would be that world capitalism would do its best to kill it, as with the other two.

ALTERNATIVE OF MASSIVE MIGRATION

Both data and theory should have convinced us by now that at least the basic-needs project, and also the equalization project, has as its condition the *equity project* (Galtung, 1980 Chapter 2*). The situation being as it is, with massive inequity, the only alternative to basic-needs satisfaction, and possible equalization, is massive migration. Nomadism was also massive migration — usually cyclical in the sense of coming back to the point of departure, and making every point on the cycle both arrival and departure. There was a move when the resources available to the population, R/P, became too low. As people could not generate more resources with the available technology, they went where the resources were available, or reduced P through infanticide, leaving elders behind to die, or splitting up; these are techniques which are also used today in sedentary societies. The major difference of today's world from former times is that now there is somebody or something wherever you move — if not people, then their domestic animals or at least a border, and behind that the coercive power of a state, wielded by police officials, or military.

According to R/P logic, population pressure on scarce resources will always remain a major factor in directing the flow of mass-migration. The hypotheses to be derived from this are obvious:

[H1] from overpopulated to underpopulated regions
[H2] from resource-poor to resource-rich regions
[H3] from high to low population-growth regions.

As a first exploration in more concrete terms, let us define 'overpopulated' as having a higher percentage of the world population than of the world's productive land territory, and 'under-populated' as having a lower percentage of the same asset. Countries can then be ranked as to how much more or less population they have, the top ten of each category being indicated in Table I.

---

* Explaining the difference between inequality (people do not have equally much) and inequity (structurally-induced inequality).

TABLE I

Top Ten Overpopulated and Ten Underpopulated Countries. *

Overpopulated			Underpopulated		
[O1]	China	+14.32	[U1]	ex-Soviet Union	− 10.78
[O2]	India	+13.38	[U2]	Canada	− 6.83
[O3]	Japan	+ 2.19	[U3]	Australia	− 5.33
[O4]	Bangladesh	+ 1.97	[U4]	Brazil	− 3.41
[O5]	Indonesia	+ 1.90	[U5]	USA	− 1.98
[O6]	Pakistan	+ 1.48	[U6]	Greenland	− 1.60
[O7]	Nigeria	+ 1.33	[U7]	Sudan	− 1.43
[O8]	Germany	+ 1.27	[U8]	Argentina	− 1.40
[O9]	Vietnam	+ 1.02	[U9]	Saudi Arabia	− 1.31
[O10]	UK	+ 0.97	[U10]	Algeria	− 1.29

* I am indebted to S. P. Udayakumar, University of Hawaii, for his assistance in preparing the data for this table, and also to Jac-Bong Lee for similar calculations with arable land only. But total area is preferred, as land can be used in many ways.

PROBLEMS OF ENTERING U FROM O COUNTRIES

If we eliminate U6–U10 because of too much ice, sand, or unpro-ductive pampas, we are left with two lists of countries and an obvious idea as to who will aspire to migrate where — from any O to a neighbouring U country.

But the problem is, still, how to get in. The four obvious answers are parts of the elementary geopolitics of tomorrow, and are as fol-lows:

First, counteraction of the coercive power of 'the other' state by your own coercive power; in other words *conquest*. This is what the most expansionist nations in the contemporary world, the colonial powers of the Occident and Japan, did to reduce their population pressure — the former for five hundred years since AD 1492, the latter for fifty years since following the start of the Sino–Japanese war of AD 1894–95.

Second, as *refugees*, non-violently seeking permanence in the recipient country. The reasons to become a refugee can be under-stood in terms of basic needs' categories: (1) escaping from being killed, seeking survival; (2) escaping from being repressed in a country that is massively breaking human rights' rules, thus seeking freedom; (3) escaping misery, including that from ecocatastrophes, hence seeking well-being; (4) escaping alienation while seeking identity. The first two are usually called *political refugees* and the third category either *economic* or *environmental refugees* according to their reasons for emigrating. The fourth category may be called *cultural refugees*, as when Muslims escape secular regimes or others escape Muslims. The needs themselves transcend the material and the somatic; they also tend to be non-material and spiritual.

Third, through *differential fertility*. The beginning is a trickle of individual migrants, even legal ones, who then start procreating, gradually outnumbering the locals, or at least so it is feared. The rest is a problem of type of regime, and democracies, equating outnumbering with overpowering through elections, would be particularly vulnerable to this approach.

Fourth, by *weakening the country of arrival*. This is much broader than the primitivism of military conquest, or weakening another country through killing and maiming or the threat thereof, in other words by means of direct violence.

Again we may proceed by basic needs' categories:

Economic weakening is what happens to people all over the world in its economic periphery, as massive structural violence, including that in the United States, gradually reduces their livelihood until the people ask for a massive wave of immigrants to bring in capital for investment, or even new blood for 'renewal of the race'. A special case is the massive introduction of disease by Western colonizers, hopefully as acts of omission rather than intended commission, but doing little or nothing to stop these 'Acts of God' as they have been called (such as doing little or nothing to counteract AIDS in Africa).

Political weakening is what happened in the Soviet Union, partly caused from the outside by exposing them to such massive arms-race pressure that authoritarian/totalitarian regimes were reinforced and democratic processes repressed as countermeasures. The result today is a virtual takeover by foreign capital and/or domestic mafia.

Cultural weakening takes place when the sense of meaning vanishes and people resort to alcohol, drugs, etc. This technique was used in the past by some Western colonizers, and the result is still visible in the alcoholism, criminality, and drug-addiction, of formerly-colonized peoples. The technique can also be used against former colonizers, as when Sendero Luminoso cooperated with Andean drug producers and distributors, knowing the insatiable US demand within large layers of the population, thoroughly alienated by the system.

DIRECTIONS OF MASSIVE MIGRATIONS

The above account gives us some key hypotheses for predicting the general direction of massive migration, spelling out three aspects of livelihood:

[H4] from low to high human-rights implementation regions
[H5] from low to high economic well-being regions
[H6] from low to high cultural-identity regions.

It should be pointed out that economic well-being is not the same as economic growth. A country may be high on the former and low on the latter and yet very attractive, e.g. USA. Another country may be low on the former and high on the latter, and still very unattractive, e.g. China. Livelihood is derived from economic level, not from growth alone. Also, one may be attracted by past dreams, such as the 'American Dream', even if they are unrealistic — rather than by present non-dream realities.

The cultural identity factor is particularly important for minorities in diaspora*, especially when such dispersion involves yearning for the 'mother country'. The Jews would be one example; the Palestinians could one day become another, and so could millions of the inhabitants of ex-Yugoslavia and the former Soviet Union.

But there is another aspect to the cultural identity thesis, because history has also defined 'mother country' as the colonial country. However much colonialism has been, and is, resented, the colonial metropolis is still highly attractive. A centre–periphery system is centripetal, not only centrifugal, and this also holds for migration. Colonialism was asymmetric, giving the colonized more competence, including linguistic, and more knowledge about the 'mother country' than *vice versa*. So people flow to the centre, sometimes with a vengeance, as if chanting *'now it's our turn!'*

There is nothing strange about this, when the 'mother country', wherever it may be, has spent so much energy and resources on legitimizing its stronghold on its colonies. Moreover, much has been done to make it look as if being colonized — meaning militarily occupied, politically dominated, economically exploited, and in addition culturally brainwashed — is close to an honour for the colony. Independence comes: the troops go home, dominance in its most abject form is over, and economic exploitation takes on new and (slightly) more subtle forms. But the cultural (especially linguistic) brainwashing remains and is reinforced by the media and 'development assistance'. Whether colonizer or developed, in either case we have a role model for states, nations, and people — with, as a result, more or less massive migration heading for that role model.

Having made use of the first two hypotheses H1 and H2, for Table I, we can now explore the impact of the factors in the other four hypotheses on massive migration flows in years to come. For this an image of the geographical world is needed, in order to locate countries or regions of geographically differential population growth, human-rights' fulfilment, economic achievement, and cultural factors.

---

* *i.e.* which are being dispersed. — Ed.

The image used here is primarily race/culture-oriented. But as there is considerable correlation between culture and geography in terms of North and South, West and East — and more particularly in terms of Northwest, Northeast, Southwest, and Southeast — geography in the 'compass sense' also enters.

The borderline between 'North' and 'South', as conceived of here, passes between the US and Mexico, runs through the Mediterranean and the Black Seas, and then follows the borders of the former Soviet Union; and the borderline between 'West' and 'East' puts the Slavonic peoples in the East and separates the Muslims and the Hindus from the Buddhist peoples in Asia. We get:

TABLE II

The Four Corners of the World.

	WEST	EAST
NORTH	*I. Northwest* North America Western Europe **USA EU**	*II. Northeast* Eastern Europe ex-Soviet Union **Russia +, Turkey +, CIS**
SOUTH	*III. Southwest* Latin America, Caribbean Arab world, Africa West Asia, South Asia **INDIA**	*IV. Southeast* Southeast Asia East Asia Pacific Islands **CHINA JAPAN**

The 'superpowers' are indicated in boldface in Table II (EU = European Union; Russia + stands for Slavonic-Orthodox countries; Turkey + stands for Turkey and Muslim ex-Soviet Republics, possibly Iran, Pakistan, Afghanistan, and Iran, all of them non-Arab). The Cold War was between Northwest and Northeast; the 'North– South' conflict was between Northwest and Southwest ('Third World'). There is also the struggle between the two growth-centres, Northwest and Southeast, for market supremacy everywhere, and over the rest, the new Periphery of world capitalism: Southwest + Northeast.

RESULTANT POPULATION PRESSURES

The North has some 20% of the world's 5.8 thousand millions of human population and is predominantly white. Apart from Muslim countries, the population growth is below the point of replacement. The white minority is decreasing in numbers. The South has 80% of

  the world's population and far above replacement growth, altogether
leading to a general pressure of South on North (H3 above).

  The political democracies are in the Northwestern corner, leading
to extra pressure (H4 above), and the economic well-being is in the
Northwestern and Southeastern corners, where the former colonial
powers are located, also leading to additional pressures.

  Even though race has been proven again and again to be a factor
of secondary importance, other factors being equal, it is bound to be
very significant, being by definition visible, and consequently
serving as a conflict-enhancer. In fact, looking at the six hypotheses
for mass-migration, all six factors correlate, and even highly so, with
race — particularly if we include the idea of the former colonial
country as a motherland in connection with cultural identity. In
practice this means that any migratory move, or indication in that
direction, will be seen in a 'bio-racial light'.

  In addition to the above comes the ethnic/cultural factor: Whites,
predominantly Christian, Muslim, or Jewish; Yellow, Buddhist–
Confucian; Brown, Hindu–Muslim; and Black and Red, held to be
animist etc. even if in practice they are Muslim or Christian. One
image is given in Fig. 1.

  The remaining 6.4% are mixed or unclassified. Worst treated (by
the Whites) in the past were the Red (through extermination, including genocide) and the Black (through slavery). White rule was based
on treating the other two somewhat better, being in an upper-middle
class alliance of nearly 81% rather than in a vulnerable minority of
23% or less. But that alliance could be fragile; the Yellow–Brown
might any day see their future against Whites, rather than with them.

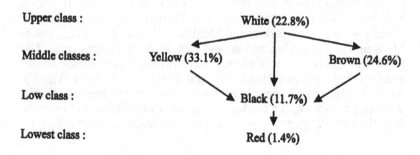

FIG. 1. *The racial composition of the world.**

---

  * I am indebted to S.P. Udayakumar, University of Hawaii, for his assistance in
preparing the data for this figure.

GEOPOLITICAL REALITIES AND ALTERNATIVES

These are the stark geopolitical realities underlying the dark altern-ative to massive migration or massive development: massive killing, or at least the systematic demopolitics of selective 'family planning', reducing the relative numbers of non-whites. This becom꞉s even more clear when we consider the total *problématique* in a more historical perspective (Fig. 2).

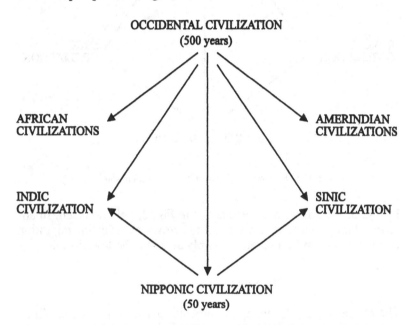

FIG. 2. *Expansionist civilizations as centrifugal forces.*

The Christian Occident expanded with devastating effects for African and Amerindian civilizations (today the latters' growth-rates are the lowest in the world); somewhat less devastating for the Indic and Sinic civilizations; Japan was left almost untouched (but then had to be 'taught a lesson' in an all-out war). Japan expanded in the Sinic sphere and in Asia–Pacific in general, including the effort to reach India. In addition the Christian and Jewish Occident en-croached upon the Muslim Occident, with the Jewish colonization of Palestine still going on.

Let us now reverse the picture, in line with H6:

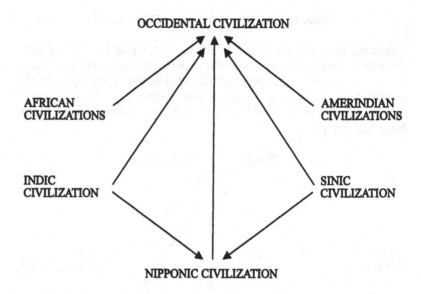

FIG. 3. *History in reverse: centripetal force with a vengeance.*

In Fig. 3, the arrows are the same as in Fig. 2, only reversed. In addition there would be corresponding arrows for Muslim migration into Christian lands, and later possibly also into the Jewish state.

RECENT ETHNIC INVASIONS

The clearest examples today are probably African, Indic, and Sinic, migration into England; African and Muslim African (Arab) migration into France; Amerindian ('Chicano') migration into the US (also in response to neo-colonialism), and Korean migration into Japan. The patterns in England and France can be described as time-zone occupation. Parts of England look very 'white' in day-time, but from 10 pm (22 hrs) until 7 am the coloured migrants take over, as service people, and are seen all over the place. That time-zone might very well expand from 9 hours to 12 hours some years hence. There is also an obvious spatial aspect, with northerners disliking blacks* who seem to hold the numerical ascendency in parts of the south.

---

* In many cases (including our own) it may not be so much a matter of liking or disliking as of deploring their invasion of already overcrowded parts of northern countries to get their social-service and other advantages without earning them. Yet when we were founding dean of the largest faculty of an elite Nigerian university we were warned that, if any of our non-Nigerian colleagues for any reason became jobless, they would have the right (or did they say 'be allowed'?) to stay in the country no more than 48 hours. — Ed.

Summarizing, here is an image for the next 500 years:

TABLE III

Six Migration Factors Impacting on Four World Corners.

	NORTHWEST	NORTHEAST	SOUTHWEST	SOUTHEAST
H1: Population density	+ US, Canada	+ ex-USSR	– Brazil	– Australia
H2: Resource density	+ US, Canada	+ ex-USSR	– Brazil	– Australia
H3: Population growth	+	+	–	–
H4: Human Rights implementation	+	–	–	– Japan, Austr.–NZ
H5: Economic well-being	+	–	–	+
H6: Cultural identity	+	–	–	+
Total of plusses.	6	3	0	2

The conclusion is clear: massive pressure on the world's Northwest, and not only from the world's Southwest (from Latin America and the Caribbean on US and Canada; from former colonies on the EU countries), but also from the world's Northeast (particularly on Germany and other EU countries).

There will be some pressure on the world's Northeast because of the low population density in ex-Soviet Union countries (particularly on the CIS; the Muslim countries are more Southwest).

There will be no pressure on the Southwest, except for half-empty Brazil. This is where the pressure originates.

There will be some pressure on the Southeast: Australia (and New Zealand) for R/P reasons, and on Japan for the last three reasons (H4–H6); above all from the Third World inside the largely rich Southeast (high-growth China, Indonesia, Thailand, the Philippines, and the Pacific Islands); and from the former colonies of Japan in Asia–Pacific.

COUNTERACTING MIGRATION PRESSURES

How will countries counteract migration pressure — apart from closing their borders, opposing migrants with police and military violence, *ius sanguinis* instead of *ius solis*, and some scattered development and demopolitical projects? And with Japan and Australia–New Zealand using geographical isolation: easier to stop swimmers and flyers than crawlers on the ground — unless giant submarines land on lonely coasts in dark nights?

One formula would be to reverse the above six factors to make countries less attractive than they currently may seem. Thus, they might stubbornly deny all rumours about available resources, making R/P look low, rather than high. They might communicate, rather than conceal, cases of human rights abuses against migrants as a deterrent, even encouraging racism (difficult in USA, which is comparatively tolerant).

Countries might also make it clear that migrants enter at the end of the line, and that ahead of them are 20%–30% of the population who are also destitute (as in the USA). Entry points ahead of others would only be for PhDs or, better, holders of senior doctorates, or people with substantial investment capital!

And when it comes to the cultural factor, countries should make it equally clear that would-be immigrants cannot have it both ways: either they stay at home and enjoy their own culture, or if they emigrate it must be on the terms of the country which they seek to enter, with due adoption of its culture, basically to serve its purposes and without expecting such nonsense as special-language schools and other preferential treatment for immigrants!

Massive migration, massive development, massive killing: the choice is ours. To believe that present flagrant misery and inequality must continue and even become worse, is not only immoral but also plumb stupid.

REFERENCES

GALTUNG, JOHAN (1980). *The True Worlds.* The Free Press, New York, NY, USA: xii + 469 pp.

The present chapter is widely derived from the above book, though additional items dealt with here will be treated in more detail in a forthcoming book and consequently are not referenced separately.

10

Women and the
Family Planning Imperative

by

PRAMILLA SENANAYAKE, MBBS (Colombo), DTPH,
PhD (London)
Assistant Secretary-General,
International Planned Parenthood Federation,
Regent's College,
Regent's Park,
London NW1 4NS,
England, UK.

INTRODUCTION

Women's roles in society, particularly their contribution to family well-being and social stability, are widely conditioned by their ability to control their own fertility.* Family planning is the responsibility of couples, not of women alone. But fertility is the ultimate gender-issue, because it is one of the few conditions in human societies in which the roles of men and women cannot be reversed. Because of her reproductive role, which is biological, a woman bears the main burden of childbirth. It is women, therefore, who suffer most from uncontrolled pregnancy and childbirth and, conversely, whose motivation is highest when decisions to space and limit births are taken (Smyke, 1991).

Women make up half the population of the world. In many countries, the best years of their lives, from adolescence to the menopause, may be squandered when they are denied the knowledge and means to control their own fertility.* For the world as a whole, this represents a loss of human resources on a massive scale. It is a paradox that while, on one hand, human numbers escalate, on the other, the potential of and for those already born is so severely and unnecessarily diminished.

---

* Here measured by the number of children to which a woman gives birth. — Ed.

185

REPRODUCTIVE HEALTH

The reproductive health of poor women in the less developed[†] and developing countries is appalling. Maternal mortality- and morbidity-rates in them range from 10 to 100 times as high as they are for women in developed countries, which constitutes the biggest differential in any socio-economic measurement contrasting developed and developing societies. In developing countries, some 120 million pregnancies occur each year, at least 40 million of which are unplanned and/or unwanted. An estimated 500,000 women die each year from pregnancy-related causes, and another 100 millions suffer morbidities or long-term disabling diseases.

At these rates, unwanted and high-risk pregnancies are killing more than 1,300 women world-wide every day (Tinker & Koblinsky, 1993). Obviously, therefore, a woman's ability to control her fertility is a major element in allowing her to improve her own and her family's prospects for the future, including her chances of participating in initiatives for development and global security. If women are not freed from the excessive health-risks which they bear, one of the greatest resources for environmentally sustainable development will be wasted (United Nations, 1992).

The maternal mortality-rate is today the most predominant indicator for reporting on women's health in developing countries. WHO (World Health Organization, 1992) defines maternal death as the death of a woman whilst pregnant or within 42 days of delivery or termination of pregnancy, from any cause related to or aggravated by the pregnancy or its management. Most direct obstetric deaths are caused by haemorrhage, infections, toxaemia, obstructed labour, or complications from spontaneous or induced abortion. These account for about three-quarters of maternal deaths in the Third World; if anaemia, which is an indirect cause, is included, the proportion rises to about 80% (AbouZahr & Royston, 1991).

Some fifteen million children in developing countries die every year without reaching the age of five years. One of the major causes of child death is the mother's poor 'reproductive health', which reduces the child's chances of survival. If a mother is malnourished, her child is likely to be below normal birth-weight, with a consequent far greater chance of being weak and at risk of contracting illnesses resulting in chronic ill-health or death. Of the nearly 130 million infants born each year, 20 millions (16%) have a low birth-weight; 95% of such infants are born in developing countries.

---

† Hereinafter included in the vague category of 'developing'. — Ed.

Where total fertility-rates are high, women tend to experience their first pregnancy early in life, which increases the risks associated with first births. High total fertility-rates also often coincide with continued child-bearing into older ages, as well as with the fact that larger total numbers of births per woman will usually mean shorter intervals between pregnancies. Each of these aspects implies increased risks to maternal health and also to infant and child health.

Sexually-transmitted diseases (STDs) continue to occur at unacceptably high levels all over the world. Urbanization, unemployment, economic hardship, and a relaxation of traditional restraints on sexual activity, have been blamed collectively for the intransigence of this situation, as also has the emergence of antibiotic-resistant strains of pathogenic microorganisms.

This growing problem of increasing STDs is nowhere more evident than in the struggle against AIDS (Acquired Immune Deficiency Syndrome) — a struggle which is particularly relevant for women in the developing world. For these women, AIDS is a scourge that exploits their poverty, their lack of proper health-care, their higher-than-normal risk of infection, and ultimately their lack of power to control their lives.

In addition, it should be noted that sexually-transmitted diseases in general affect women more than men — especially through their serious sequelae, which include pelvic inflammatory disease, infertility, ectopic pregnancy (*i.e.* outside the womb), and cancer of the cervix. The risk of transmission is greater from man to woman than from woman to man. In addition, women can transmit some diseases to the foetus.

ADOLESCENT HEALTH

For female adolescents, having children at an early age presents serious health and social problems. Their future educational, employment, and social, opportunities may be severely curtailed. Teenage pregnancy is a serious health-problem in many countries. The sheer size of the adolescent population indicates the scale of need for their reproductive health-care. In 1995, according to the World Bank (1993), there were expected to be 512 million young people aged 15 to 19, 426 million of them (83%) living in developing countries. These numbers are expected to continue to grow.

In both developed and developing countries, early child-bearing increases the chances of health complications. Adolescent mothers may run double the risk of dying during childbirth as compared with women in their early 'twenties. In many developing countries, complications of pregnancy, childbirth, and the puerperium, are among the main causes of death in 15–19 years-old girls.

The death-rate from causes related to abortion is particularly high in girls aged 15 to 19, who account for at least 10% of the estimated 50 million induced abortions that occur each year throughout the world. A study in Benin City, Nigeria, reported that 72% of the deaths of young women were related to complications of unsafe abortions (UNDIESA, 1989). There are also other dangerous practices linked to the prevailing moral norms and inappropriate laws in certain countries. For instance, in all the Arab-world countries except Tunisia, abortion is illegal and unwed mothers are unacceptable to society. Each year, some 2,000 children are abandoned by their young unwed mothers in Algeria and Morocco alone, while reports abound of abandoned children in the streets in Egypt. In a survey carried out in Algeria between 1979 and 1984, it was shown that 30% of women who committed suicide during this period were pregnant and unmarried (IPPF, 1992).

Although adolescents are becoming sexually active earlier and earlier, their contraceptive practice is very low. In Benin, Cameroon, Côte d'Ivoire, Kenya, and Nigeria, between one-fifth and one-third of teenage females had premarital conceptions (UNDIESA, 1989). Because of the severe restrictions on access to contraception and information for young people, one of the most popular methods is the rhythm method — with, however, a very poor knowledge of the fertile period.

In 8 of 11 countries investigated in Africa, more than half of adolescent women knew of at least one modern method of contraception, but few had actually practised any. The best rates of knowledge and use were in Botswana; the worst in Mali.

In both developed and developing countries, too many women become mothers at too early an age for their own welfare and that of their children. Early marriage and child-bearing are closely linked to high total fertility. Girls who marry between the ages of 15 and 19 are likely to bear, on average, six or seven children.

Births to women under 20 years of age represent an increasing proportion of all births — a fact that is partially explained by the comparatively greater number of young people in the populations of developing than developed countries. The number of young people under the age of 15 in the population of the world as a whole grew from 700 millions in 1950 to 1.7 thousand millions in 1990. The highest rates of increase occur in the developing countries, with 40% of the population under the age of 15 in East, Middle, and West, Africa and 39% in South Asia (United Nations, 1992). Such figures indicate the trend in numbers of young women entering reproductive age.

The scourge of teenage pregnancy is not confined to developing countries. In the United States, according to recent estimates by the American Enterprise Institute, a conservative think-tank, more than

1 million adolescents are likely to become pregnant in any year, resulting in 400,000 abortions, 134,000 miscarriages, and 490,000 actual births — two-thirds of them out of wedlock. Globally, more than 15 million girls aged 15 to 19 give birth every year, while some 5 million others have abortions (ICAF, 1992).

Patterns of sexual behaviour in many human societies are being altered by biological, social, and economic, factors. Among these, early sexual maturity, changing moral values, and later marriage, figure prominently. Migration from village to town or from one country to another, and the influence of the mass media, have in some countries lessened the influence of social mechanisms which formerly discouraged adolescent sexual activity before marriage. Presumably as a consequence, increasing numbers of cases of sexually-transmitted diseases are being reported among young people.

Both traditional and modern societies are increasingly apt to expose very young women to unwanted pregnancy, which can ruin their entire lives by cutting off for ever their chances of education, employment, and personal self-fulfilment. It is essential that young people of both sexes should be carefully prepared for responsible parenthood through appropriate sex and family-life education in and out of school. Services for young women are needed which cater to their special needs and protect them from exploitation (Senanayake & Ladjali, in press).

FAMILY PLANNING — THE ULTIMATE IMPERATIVE

Fertility regulation is central to all other aspects of reproductive health. It contributes to the prevention of sexually-transmitted diseases, to alleviation of the consequences of unwanted pregnancy,* infertility, and sexuality, while contributing to child survival and safe motherhood. Wise family planning accordingly saves the lives of women and children and, by reducing family size, hastens the 'demographic transition' to population stabilization. Above all, it gives women control over the unique part of their own bodies such as otherwise eludes them.

In industrialized countries with stable or declining populations, modern contraception is widely practised, a range of contraceptive options being readily available. A revolution in contraceptive tech-

---

*Here we cannot help relaying our Wife's long-term contention that a large proportion of the world's worst and cruellest dictators seem to have been unwanted children, born against the wishes of their parents or anyway mother and doubtless treated to some extent accordingly. — Ed.

nology in the last few decades has produced safe and effective methods, of which 'The Pill' and the intrauterine device are the most widely used, though injectables and sub-dermal implants are gaining in popularity. Condoms, while less effective than hormonal contraceptives, are the main method for men and popular as a prophylactic. Both male and female sterilization is common for those who have completed their desired family-size.

Many millions of couples and individuals, however, still do not have access to family planning — either because they lack information or because appropriate methods, counselling, and follow-up services, are not available. Data from the World Fertility Survey revealed that 300 million couples who wanted no more children were not practising modern contraception (UNFPA, 1991); the true figure is probably more than 500 millions when account is taken of the large numbers who are sexually active but unmarried, who seek abortions, or who have had to give up practising contraception because of dissatisfaction with the method they were offered or the absence of sympathetic and effective counselling, advice, and clinical services.

In many developing countries, family planning services are still not widely available — mainly due to lethargic government attitudes to the provision of health-care in general, to Catholic opposition, and/or to ignorance about the social and economic benefits to be derived from its promotion. Lack of services correlates with high rates of population growth, high maternal and child mortality, poverty, and the rapid depletion or degradation of natural resources.

Nevertheless, there are signs of improvement. Use of modern contraception rose from less than 10% in the 1960s to 45% in 1983, and to 51% in 1992 (United Nations, 1992). The total fertility rate — the average number of children per woman — in developing countries has declined from 6.1 in 1965–70 to 3.9 in 1985–90, with, however, great regional and national differences. The speed of decline in some parts of the world is striking. It took 58 years for fertility in the United States to decline from 6.5 to 3.5, while it has taken Thailand only 8 years and China only 7 years to achieve practically the same result (UNDIESA, 1989).

The statistics on induced abortions provide another indication of the level of the unmet need for family planning in developing countries. It is estimated that between 36 and 53 million induced abortions are performed each year — an annual rate of 32–46 abortions per 1,000 women of reproductive age. Many of these abortions are clandestine, in countries with repressive legislation, and so the exact number is impossible to determine. But it is estimated that, out of the 500,000 maternal deaths which occur each year throughout the world, as many as from one-quarter to one-third may be the result of complications from unsafe abortion procedures (Fathalla, 1992).

In developing countries, statistics on abortions are sketchy. For example, in India about 600,000 legal abortions are recorded each year in government hospitals but it is thought that at least another four millions take place illegally outside these institutions. In Bangladesh the law permits abortion only to save a woman's life, but menstrual regulation by vacuum aspiration is allowed soon after a missed period without testing for pregnancy (McLaurin et al., 1991). Surveys indicate that perhaps a quarter of a million legal terminations or cases of such menstrual regulation occur, but there could be three times this many illegal abortions. An estimated four million illegal abortions occur each year in Latin America (Henshaw & Morrow, 1990). In Africa and the Middle East, where large families are still considered desirable, it has been assumed that abortion was more rare; but recent studies, for example in Kenya and Nigeria, have shown that abortion complications are responsible for a large percentage of hospital admissions and pregnancy-related deaths (Senanayake, 1992). An estimation of abortion prevalence in Kenya (Kenya National Council for Population and Development, 1989) gives a figure of 75,000 induced abortions in that country. Extrapolation for sub-Saharan Africa arrives at an estimate of 1.5 million abortions annually in that region.

The World Health Organization (WHO) estimates that more than half the deaths caused by induced abortion occur in South and South-east Asia, followed by sub-Saharan Africa. In all parts of the world a small, but increasing, proportion of abortion-seekers are unmarried adolescents; in some urban centres in Africa they represent the majority (ICAF, 1992).

There is indisputable evidence to show that the risks associated with child-bearing are considerably reduced when births are adequately spaced — starting not too early, and ending not too late. Family planning also saves children's lives by enabling a proper interval between the birth of siblings. One out of five infant deaths in developing countries — there are more than eight million infant deaths annually (i.e. under one year of age) — would be prevented if all births were spaced by an interval of at least two years (UNICEF, 1993).

The contribution of family planning to saving the lives of infants and children is thus of great significance. Contraceptive use has both direct and indirect benefits because it:

— helps very young women whose infants are prone to higher mortality to delay child-bearing until a later age;
— allows older and especially high-parity women*, whose infants are at considerably higher risk of dying, to stop having babies;

---

* i.e. who have already borne many children.

— contributes to longer intervals between births, which improve
  infant and child survival by allowing time for the restoration of
  health to the mother;
— reduces maternal mortality;
— changes the environment in which couples set their family size-
  goals; and
— provides a cost-effective route to reducing infant and child
  mortality.

Family planning programmes have understandably focused on
women, as the risks to their health are so high and their independence is so compromised by repeated child-bearing. But the reality
of women's lives shows that their partners have considerable influence and control in determining whether or not a woman is allowed
to use a contraceptive. A woman's access to contraceptive services
is often dependent upon the consent of her partner, who may have
little knowledge of the benefits to her and her children, or may see
threats to his own masculinity. In these circumstances, a woman may
have little chance of changing entrenched male attitudes.

The history of the family planning movement is quite recent. The
International Planned Parenthood Federation (IPPF), its leading
pioneer world-wide, celebrated its 40th anniversary in 1992. The
United Nations General Assembly, after much controversy, finally
agreed to meet requests for *ad hoc* assistance from developing countries only in 1962. The World Health Organization (WHO) established a Human Reproduction Unit in 1965, and the first United
Nations' population fund was established in 1967. This was the
precursor of what became the United Nations Fund for Population
Activities (UNFPA) and is now the United Nations Population Fund,
though widely using the same familiar acronym.

More recent advances at the international level have been the
commitment of the World Bank and UNICEF (United Nations
Children's Fund). Much of the stimulus for the involvement of these
agencies has come from the donor community, whose member
governments finance family planning activities as part of their
development assistance. In the past few decades, numerous actions
and pronouncements of the international community have supported
the universal provision of family planning services and called for
increased financial support for them.

In May 1992 a Technical Discussion on Women, Health, and
Development, convened by the World Health Assembly, concluded:

'If women are to realize their full potential in their productive
roles, they must be able to manage their reproductive role.
This means that they must have access to family planning

information and services. These are essential if women's reproductive rights are to be secured. Moreover, practices carried [on] without women's informed consent should be identified and eliminated. The ability to decide, freely and in an informed manner, the number and spacing of one's children, is the first step to enabling women to exercise other choices. When a woman realizes that she can make decisions regarding her reproductive function, this experience of autonomy spreads to other aspects of her life. It enables her to pursue diverse opportunities and empowers her to make pivotal decisions in her own life. Family planning is an essential means of enhancing women's autonomy.' (WHO, 1992).

Family planning services are a very sensitive part of maternal and child health-care. They must respect cultural and personal attitudes and include sympathetic, well-informed counselling and advice. The methods offered to clients must be those that have been rigorously tested and pronounced safe and effective, free from as many side-effects as possible, and convenient and not unpleasant to use. Fortunately, there are now many authoritative bodies, such as the World Health Organization and national drug regulatory agencies, that monitor the safety and efficacy of contraceptives. There are also effective bodies, such as IPPF's International Medical Advisory Panel, which weigh the evidence and provide guidelines on purchase, distribution, and clinical use, of various methods of family planning.

It has to be acknowledged, however, that not all services maintain the desired standards. On the whole, services provided by the private sector, including voluntary organizations, are generally of a better quality than those provided in large-scale government programmes. A frequent fault is to limit the options available to clients; a 'cafeteria' of reliable methods should be on offer, to meet the needs of different women at different times in their reproductive life.

Needless medical barriers and poor contraceptive images continue to plague the family planning service delivery system (Shelton et al., 1992). These involve inappropriate eligibility criteria, including age and parity; unwarranted contra-indications; unnecessary process hurdles, including laboratory tests; provider bias; restrictions on providers, and regulatory barriers. Reducing these medical and other barriers, and promoting a positive contraceptive image, are keys to improving services' delivery.

In its efforts to provide models for the provision of family planning services, IPPF gives priority to the rights of the client. It identifies these rights as: (1) the right to information — about why, how, and where, to obtain family planning advice and services; (2) the right of access — regardless of marital status, culture, religion, or other common barriers; (3) the right of choice — whether or not to practice or discontinue contraception and what method to use; (4) the right to safety — protection against health risks and negative effects; (5) the right to privacy — in counselling and clinical procedures; (6) the right to confidentiality — as protected under the Hippocratic Oath; (7) the right to dignity — including courtesy and consideration; (8) the right to comfort — a pleasant and comfortable environment; (9) the right to continuity — services and supplies as long as needed; and (10) the right to opinion — especially on the quality of services and how to improve them.*

The revolution in contraceptive technology, which produced today's modern methods, lost its momentum after the first wave of significant achievements. Research into new methods — especially male methods — is a neglected area at present. The multinational pharmaceutical companies have largely withdrawn from this work because of the risks involved, the low return on investment, and the lengthy process of getting the approval of drug-regulatory authorities to put new items on the market. Other factors are the failure of the donor community to maintain a high level of support for contraceptive research, and the scientific community's lack of progress in exploiting new advances in cell- and molecular-biology and biotechnology for contraceptive purposes.

There is a critical need to reverse the above trends in contraceptive research and technology, to enable more and better methods to be developed — especially methods which can at the same time help to prevent the spread of sexually-transmitted diseases (Fathalla, 1993). This was recognized by an International Symposium on Contraceptive Research and Development for the Year 2000 and Beyond, held in Mexico in 1993. In a declaration, it called for more and better scientific research facilites, greater financial support, increased involvement of industry, and the collaboration of national drug and device regulatory agencies.

---

* It has long seemed to us that a serious impediment to full agreement on the need for solid advance towards the ideal of 'universal' family planning could be the taboo nature of the entire subject of sexuality and procreation very widely in the decision-making echelons of human society — even to the extent that will-educated moderns do not have due patience with the reticence that older people, who have been 'properly brought-up', have about discussing such topics, at least at all openly. Is this, perhaps, behind the difficulty we have encountered in getting modern editors (*e.g.* of *Science*) and others to unterstand our conviction that a major factor in questions of procreation can and should be human self-discipline and concomitant physical restraint? — Ed.

INTERACTION BETWEEN FAMILY PLANNING AND THE STATUS OF WOMEN

The links between unwanted pregnancy and other forms of discrimination against women have been extensively researched and catalogued in the past few decades. Poverty, poor health, and lack of education and of paid employment, are all implicated; they also perpetuate the dismal social conditions that are often at the root of conflict. The widespread neglect of women's health contradicts sharply the growing perception of women as agents of change.

Discrimination against women perpetuates their high fertility which, in turn, limits women's options for any activity outside the home. It has negative consequences for the conduct of their lives in virtually all after-spheres, and is the most formidable obstacle to their participation in most forms of development. The empowerment of women — their ability to control their own lives — is the foundation for all action linking population, environment, development, and global security.

Many taboos against women are enshrined in laws that legitimize the inequality of the sexes. Laws also restrict women's access to medical services where, for example, married women are required to obtain their husbands' permission before receiving contraceptive services or undergoing voluntary sterilization. Abortion laws are a case in point. Women resort to abortion everywhere, despite restrictive laws and/or religious opposition. Only 23% of the world's population live in countries where the law permits abortion for social and social-medical reasons, while 25% of the world's population live in countries where abortions are available only if the life of the woman is in danger.

The safety and legality of abortion are matters of life and death for women. Contraception can substantially reduce, but not entirely eliminate, the need for abortion. Many people concerned with the reproductive rights of women argue that safe abortion should be recognized and promoted as a method of family planning. IPPF, reflecting the diversity of attitudes within its far-flung membership, has moved cautiously on this issue, but now recognizes the need to eliminate the high incidence of unsafe abortion and increase women's access to safe, legal abortion.

Even in countries where abortion is legal, many barriers are put in the way of access to safe abortion. Abortion services may be insufficient to meet demand, or they may be inaccessible — especially to rural women. They may cost too much and be beset by administrative delays and obstacles. Hospital regulations, or doctors who oppose abortion, may also block access to abortion services. Permission of husbands or parents may be needed. All this leads to abortions at later stages of pregnancy, with associated medical and emotional consequences.

Discrimination against women also shows up in laws regulating the age of marriage. Younger ages are fixed for women than for men, leading women into early child-bearing and denying them the extra years of education and preparation for life. Girls are often married as young as 13, and commonly by the age of 16, in many developing countries, thus ensuring a very long reproductive life with its consequent health-risks. The still-widespread absence of laws protecting women's health leaves the way open for adolescent pregnancy and high rates of maternal mortality and morbidity.

Women derive their status primarily from their child-bearing role, and their value is often measured by the number of sons they bear, preference for sons being strong in many countries. Girls are often considered an economic liability; thus in India, for example, the dowry tradition imposes a financial burden on the parents of girls. When this is added to lack of property rights and control over economic resources, it is difficult for a woman to provide financial support for ageing parents and therefore she is undervalued by them relatively to boys.

Because of the supposition that girls are a drain on family resources, families are unwilling to invest in daughters. In many developing countries, the discrimination against women is reflected in their poor health-status compared with men. From the beginning of their lives, different feeding practices, additional burdens of work, and lack of basic education, put girls at greater risk than boys of suffering malnutrition and/or contracting disease. Other practices, such as early marriage, forcing of women into the reproductive cycle before they are physically and socially mature, still widely lead to repeated pregnancies — often at the risk of the infants' and/or womens' lives (WHO, 1992).

The practice of purdah denies women access to the outside world and prevents them from forming partnerships with other women — e.g. to improve their condition and learn inter alia about human sexuality and fertility control.

Where reproductive health is concerned, religion and politics have taken precedence over public health priorities. In recent decades, for example, women in the former Soviet Union and much of eastern Europe have relied heavily on unsafe abortion because of the enforced scarcity of suitable contraceptives and safe abortion services — a policy intended to increase birth-rates. Similarly, in the United States and elsewhere, women's legal right to safe abortion services is frequently under threat (Eschen, 1992).

The opposition of the Roman Catholic Church to the use of modern contraceptives is a major stumbling-block. Many priests in

Roman Catholic countries recognize the health and economic problems which their parishioners face, and do not stand in the way of family planning programmes; and many millions of practising Catholics defy the church on contraception. But the position of the Vatican has inhibited some governments from embarking on full-scale, nation-wide family planning programmes, and has had an often disastrous effect on the international community, as when the Reagan Administration wiped out all US funding for the United Nations Population Fund (UNFPA) and the International Planned Parenthood Federation (IPPF) in 1985 and until it was restored by the Clinton Administration in 1993. The Vatican's continuing opposition to family planning has now been restated in a recent encyclical *Vitatis Splendor.**

There is a strong correlation between the education of girls and women and fertility decline; indeed education and smaller families practically go together. Poorly-educated women in Brazil, for instance, have an average of 6.5 children, but those with secondary education only 2.5.[†] In Liberia, women who have been to secondary school are 10 times more likely to take advantage of family planning facilities than those who have not been to school at all (UNFPA, 1991).

On the average, women with no education have twice as many children as those with seven or more years of schooling, while women with several years of schooling are also three times more likely to be using effective contraception. In Pakistan, where only 21% of women can read and write, women bear on average 5.9 children, whereas in Sri Lanka, despite a very similar GNP, 84% of women are literate and have on average just 2.5 children each.

It has been estimated that, in developing countries, each extra year of education for a mother reduces her children's mortality risk by an average of 7–9% (UNFPA, 1992). Better education for mothers also reduces the incidence of stunting and underweight in their children. Better-educated mothers are more likely to have had tetanus immunization, and many times more likely to use pre- and post-natal care.

---

*Here we recall how, several years ago, following an Oxford University Encaenia at which a Cardinal from Rome (who was said to be scientific adviser to the Pope) received an honorary Doctorate of Divinity. We walked back to our College with the honorand and assailed him on the uncompromising attitude of the Roman Catholic Church on this vital issue — to be assured that, whereas they were 'opposed to abortion', they [we think he said 'we' but it may have been merely 'I'] are/am 'not opposed to family planning'. Writing subsequently to the Cardinal with a plea to use his influence in that most important respect in the Vatican in the global interest, we received no reply but were later informed that 'the Cardinal has retired'. — Ed.

[†] Remarkable figures reminding us of how, for our first International Conference on Environmental Future in Finland some 25 years earlier, we had been warned to keep off the subject of family planning, as a strong Brazilian contingent was expected *inter alia* to counter the effect of the participation of the then Secretary-General of IPPF. — Ed.

While use of family planning is higher with increased education, it is also true that family planning and lower fertility help to improve women's status and education. When girls leave school early for marriage and child-bearing, as they often do in southern Asia and most of Africa, it perpetuates the cycle of low status and high fertility. Pregnancy outside marriage is a major cause of girls dropping out of school in Latin America and Africa (UNFPA, 1992).

Low status of women correlates strongly with poverty, the health risks of poverty being far greater for females than for males. The United Nations Human Development Index, which measures the extent to which people have the resources to attain a decent standard of living, shows that women have lagged behind men in every country for which data are available. Not only do women not automatically benefit from economic growth, but they may even fall further behind in its wake. Indeed, women lag behind men on virtually every indicator of social and economic status, through latterly their situation has tended to improve. In every country and at every socio-economic level, however, women still control fewer productive assets than do men.

Taking housework into account it seems safe to conclude that women everywhere work longer hours than men; but they generally earn less income despite the fact that they are responsible for meeting from 40 to 100% of a family's basic needs. And, lacking alternatives, women are more often compelled to resort to jobs that are seasonal, labour-intensive, and/or carry considerable occupational risk. As a result, poverty among females is more intractable than among males, and their health is even more vulnerable to adverse changes in social and environmental conditions (Koblinsky et al., 1992). Poverty and environmental degradation lead to a 'vicious cycle' of poor nutritional status, high infant mortality, and consequent high rates of pregnancy — especially where parents expect their offspring to take care of them in cases of old-age or infirmity.

Women's economic opportunities affect desired family-size. One analysis of the World Fertility Survey data from 20 developing countries found that female participation in the labour force was the single most important determinant of reduced marital fertility (ISI, 1984). But women without control over their own fertility have limited opportunities to enter paid employment.

Women's access to the labour market brings multiple benefits. It helps to lower fertility by delaying the age of marriage. After marriage it provides women with an independent income which will improve their power and status in the family and community. It helps children directly, because far more of women's income than men's goes into the welfare of children. Recent research shows that the proportion of children in poverty is much lower where the mother works for cash (UNFPA, 1992).

It should also be noted that there are obvious links between family planning and women's capacity for management of natural resources. Repeated pregnancies, close together, weaken the health of women, reducing their ability to fulfil their roles as resource managers. Faced with a constant struggle for daily survival, people will use natural resources differently, and with less consideration for longer-term conservation, from those people who depend on natural resources but whose basic needs for survival have been taken care of.

A growing area of concern to family planning programmes is migration, both internally and across national borders. Some studies show that fertility among migrants is low, but the reproductive health-needs of migrating women are evidently very high. It is mistakenly assumed that most people who migrate internally and internationally are men, and immigration policies generally assume that migrants are men. However, according to United Nations estimates (UNFPA, 1993), 75% of the total global refugee population may be women, and from 60 to 80% of refugee households are headed by women.

Refugees are notoriously ill-served by health and social services — particularly reproductive health services. Thus female refugees are put at increased risk of unwanted pregnancies that threaten their fragile hold on employment possibilities. Moreover, as economic incentives are widely lacking for women, this is yet another area where the potential contribution of women to environmentally sustainable development is being widely ignored.

RECENT CONFERENCE COMMITMENTS

UN International Conference on Population and Development

A source of optimism for the future emerged when the 'climate' for family planning began to change in the mid-1990s. The 1994 International Conference on Population and Development (ICPD) — the third in a series of United Nations population conferences that has been held every ten years — gave momentum to the process by moving the focus of population programmes away from setting demographic targets and towards investing in people — especially in people's health and education — as the proper basis for sustained economic growth and sustainable development (cf. Senanayake, 1995).

The ICPD was a 'landmark' conference, which emphasized women's role in development and urged governments to intensify their efforts to advance gender equality, equity, and the empowerment of women. The Programme of Action called for the full and equal participation of women in civil, cultural, economic, political,

and social, life and for the eradication of all forms of discrimination
on grounds of sex — especially discrimination against the girl child.
Education was given priority as one of the most important and
effective ways of achieving self-determination for women, and
countries were urged to provide access to education and training for
women and girls. Practices such as forced marriage, female genital
mutilation, prenatal sex selection, and the infanticide of girls babies,
were condemned.

While reiterating the long-established right of couples to decide
freely and responsibly on the number and spacing of their children,
and to have the information, education, and means, to do so, the
Conference report also sets out the right to enjoy the highest attainable standard of reproductive and sexual health, including family
planning, and said that governments must meet people's — especially women's — total span of needs in the areas of sexuality and
reproduction. They should all try to make reproductive health-care
available as soon as possible and no later than AD 2015.

As well as embracing family planning counselling, information,
education, communications, and services, reproductive health-care is
defined as including prenatal care, safe delivery, and postnatal care;
prevention and treatment of infertility; prevention of abortion and
management of the consequences of abortion; sexually transmitted
diseases' prevention and treatment; and information on sexuality,
reproductive health, and responsible parenthood. The needs of
adolescents were specifically mentioned at ICPD, as was the
importance of high-quality services. Targets were set to reduce
infant and child mortality, and to halve the 1990 level of maternal
mortality by AD 2000, and halve it again by AD 2015. The funding
required to achieve the aims outlined in the document was also
given.

ICPD was attended by representatives from 179 governments and
over 1,250 accredited nongovernmental organizations, and the
Programme of Action was unanimously adopted by governments
with only a few reservations.

The Fourth World Conference on Women (FWCW)

This event also embraced the concepts of sexual and reproductive
health and rights. It took place in 1995 in Beijing, China, and its
Platform for Action added a clarification of sexual rights which it
said included the right to decide, feely and responsibly, on all
matters related to sexuality — which should be free from any
coercion, discrimination, and violence.

As a result of these conferences, much greater emphasis than
previously is now being placed on people's rights to sexual and

reproductive health, of which IPPF's Charter on Sexual and Reproductive Rights, drawn up in 1995, represents just one initiative (IPPF, 1995). It sets out the human rights context in which IPPF carries out its mandate in the field of sexual and reproductive health, but also, most importantly, draws attention to the extent to which sexual and reproductive rights are being recognized as *human rights* by the international community in internationally adopted UN and other declarations, conventions, and covenants, including those emanating from ICPD and FWCW. Governments can thus be held to account, and violations of rights monitored.

THE FUTURE: A NEW ETHICAL FRAMEWORK?

I believe that a new ethical framework for action is needed. Such a framework must recognize that people are the primary resource for development. In this ethical framework, women and their needs should be the central focus. Practically throughout the world, it is women who are apt to bear the greater burdens of physical labour, in their productive as well as reproductive roles.

How can real and universal development take place when half the world's population suffers social, educational, and economic, discrimination, is overworked, undernourished, illiterate, and constantly at risk of unwanted pregnancies? How can families be healthy and productive when women suffer such severe additional risks to their health as are associated with pregnancy and child-bearing. It is a global disgrace that, in today's world, half-a-million women still die each year due to pregnancy-related causes. It is totally unacceptable that young girls are married off before their bodies are prepared for motherhood. How can sound development take place when women have no control or choice over their own lives? Family planning can give women the choice and the chance to take some control. *We must give women the choice.*

It would be criminal if we do not hand over to the next millennium a world in which each and every woman in every corner of the globe has been informed about, and given the means to use, family planning options. By the year 2000 we must at least double the prevalence of contraceptive use if we really want a sustainable balance between population and resources to emerge in the next century.

At present only one per cent of development assistance goes into family planning programmes — about $1 thousand millions annually. The total spending on family planning from in-country and international sources needs to double in the next few years, and to reach at least $4 thousand millions well before the end of the next century.

At the very least, we should set ourselves the following goals:

By the Year 2000

- Universal access to primary health-care/family planning services.
- Reduce infant and under-five mortality rates by one-third, or to 50 and 70, respectively, of every 1,000 live births (whichever is less).
- Reduce maternal mortality by one-half of AD 1990 levels.
- Reduce new HIV infection rates to half the current projected rate in developing countries.

By the year 2005

- Close the gender gap in secondary school education.
- Remove all programme-related and unnecessary legal, medical, clinical, and regulatory, barriers to family planning information and services.

By the year 2015

- Achieve a life expectancy at birth of more than 75 years. Countries with the highest levels of mortality should achieve life expectancy at birth in excess of 70 years.
- Provide universal primary education in all countries.
- Eliminate gender disparities by income.
- Provide universal access to the full range of safe and reliable family planning methods and related reproductive health services.
- Ensure that all pregnancies are intentional and all children are wanted.*

If all the expressed but unmet needs for family planning over the next two decades are met, there will be an increase in contraceptive use to 69% in the developing world. Donor resource-requirements to meet these goals and activities would be (in 1993 US dollars): $4.4 thousand millions in AD 2000; $4.8 thousand millions in AD 2005; $5.3 thousand millions in AD 2010; and $5.7 thousand millions in AD 2015.

Finally, it is hoped that all unwanted births and all unsafe abortion will be eliminated.

In conclusion, it should be re-emphasized that there are still few societies in the contemporary world where women's special creativities and energies in shaping and upholding duly-caring public policies — in such areas as reproductive health, children, the youth, the elderly, the environment, and the future security of the world —

---

* Hence hopefully saving the future world from the 'worst and cruellest dictators' — *see* footnote on p. 189. — Ed.

are given proper encouragement, though there has been attempted improvement of late.* This is sad because women are considered by many to have more social imagination and commitment than men, and women are apt to be more willing then men to promote far-sighted solutions with their incumbent short-term sacrifices than men, who generally tend towards short-sighted expediency.

Health-care budgets are inadequate everywhere, and need desperately to be increased. But currently available funds are often poorly allocated, being commonly weighted towards curative rather than preventive measures. Real improvements in women's health status will require far-reaching socio-economic and cultural change, extending beyond the health-care system. Because women's health is generally a direct reflection of their status, no strategy can be fully successful in the long term unless women become equal partners in social development. Systematic and fundamental changes in the nature of policies towards, and allocation of, resources to women will have to be addressed as part of a long-term strategy of change (World Bank, 1992).

Our ethical framework in The Biosphere must recognize that people are the primary resource for satisfactory life and any future development, but that the quality of their lives, and thus their ability to participate fully, necessitates bringing their numbers into balance with the Earth's finite resources.

Unless we base all we undertake on the needs of people — on providing a decent quality of life for all peoples — we risk failure. Sophisticated technology will not solve the environmental dilemmas, nor will it prove to be effective in alleviating poverty unless the physical, intellectual, social, and spiritual, energies of people everywhere are freed to contribute to democratic development. If we lose sight of this ethical framework for development, we will be liable to fail miserably.

REFERENCES

ABOUZAHR, C. & ROYSTON, E. (1991). *Maternal Mortality: a Global Factbook*. World Health Organization, Geneva, Switzerland: 17 pp.

ESCHEN, A. (1992). *Acting to Save Women's Lives: Report of the Meeting of Partners for Safe Motherhood*. World Bank, Washington, DC, USA: xii + 329 pp., illustr.

FATHALLA, M.F. (1992). *Reproductive Health in the World: Two Decades of Progress and the Challenge Ahead*. Paper presented at the International Conference on Population and Development, Expert Meeting on Population and Women, June 1992. United Nations Population Division, New York, NY, USA: [not available for checking].

*Sympathetic though we are to all attempts at this needed improvement — women often having talents that most men lack — we are bound to admit that this cause has not been helped by some unfortunate appointments of unsuitable and even unqualified women of late to key posts involving exacting demands of judgement as well as stamina. — Ed.

FATHALLA, M.F. (1993). *Mobilization of Resources for a Second Contraceptive Technology Revolution*. Rockefeller Foundation, New York, NY, USA: 17 pp.

HENSHAW, S.K. & MORROW, E. (1990). *Induced Abortion: a World Review 1990 Supplement*. Alan Guttmacher Institute, New York, NY, USA: 120 pp.

INTERNATIONAL CENTER ON ADOLESCENT FERTILITY (ICAF) (1992). *Adolescents and Unsafe Abortion in Developing Countries: a Preventable Tragedy*. ICAF, Washington, DC, USA: 68 pp., illustr.

INTERNATIONAL PLANNED PARENTHOOD FEDERATION (IPPF) (1992). *Unsafe Abortion and Sexual Health in the Arab World: the Damascus Conference*. IPPF, London, England, UK: 31 pp.

INTERNATIONAL PLANKED PARENTHOOD FEDERATION (IPPF) (1995). [Not finally checked.]

INTERNATIONAL STATISTICAL INSTITUTE (cited as ISI) (1984). *World Fertility Survey: Major Findings and Implications*. ISI, Voorburg, The Netherlands: viii + 61 pp.

KENYA NATIONAL COUNCIL FOR POPULATION AND DEVELOPMENT (1989). *Kenya Demographic and Health Survey, 1989*. Institute for Resource Development, Columbia, Maryland, USA: xxii + 158 pp.

KOBLINSKY, M.A., TIMYAN, J. & GAY, J. (Eds) (1992). *Women's Health at the Crossroads*. Population Council, New York, NY, USA: 51 pp.

MCLAURIN, K.E., HORD, C.E. & WORLF, M. (1991). *Health Systems' Role in Abortion Care: the Need for a Pro-active Approach*. International Projects Assistance Services (IPAS), Carrboro, North Carolina, USA: 34 pp.

SENANAYAKE, P. (1992). Abortion in a global perspective. *Journal of Psychosomatic Obstetrics and Gynaecology*, **13**, Supplement 1, p. 39 (Abstract).

SENANAYAKE, P. (1995). Reflections on the UN's International Conference on Population and Development. *Environmental Conservation*, **22**(1), pp. 4–5.

SENANAYAKE, P. & LADJALI, M. (in press). *Adolescent Health: Changing Needs*. World Report on Women's Health, IPPF, London, England, UK.

SHELTON, J., ANGLE, M.A. & JACOBSTEIN, R.A. (1992). Medical barriers to access to family planning. *The Lancet*, **340**, pp. 1334–5.

SMYKE, P. (1991). *Women and Health*. Zed Books, London, England, UK: ix + 182 pp., illustr.

TINKER, A. & KOBLINSKY, M.A. (1993). *Making Motherhood Safe*. World Bank, Washington, DC, USA: xv + 143 p., illustr.

UNITED NATIONS (1992). *Women 2000: Equal Rights for Women and Girls, Nr 3*. Division for the Advancement of Women, Centre for Social Development and Humanitarian Affairs, Vienna, Austria: 19 pp., illustr.

UNITED NATIONS CHILDREN'S FUND (UNICEF) (1993). *The Progress of Nations*. UNICEF, New York, NY, USA: 53 pp., illustr.

UNITED NATIONS DEPARTMENT OF INTERNATIONAL, ECONOMIC AND SOCIAL AFFAIRS (UNDIESA) (1989). *Adolescent Reproductive Behaviour: Evidence from Developing Countries*. UN, New York, NY, USA: 128 pp.

UNITED NATIONS POPULATION FUND (UNFPA) (1991). *The State of World Population 1991*. UNFPA, New York, NY, USA: 48 pp., illustr.

UNITED NATIONS POPULATION FUND (UNFPA) (1992). *The State of World Population 1992*. UNFPA, New York, NY, USA. 46 pp., illustr.

UNITED NATIONS POPULATION FUND (UNFPA) (1993). *The State of World Population 1993*. UNFPA, New York, NY, USA: 54 pp., illustr.

WORLD BANK (1992). *Development and the Environment: World Development Report 1992*. Oxford University Press, Oxford, England, UK: 308 pp., illustr.

WORLD BANK (1993). *Investing in Health: World Development Report 1993*. Oxford University Press, Oxford, England, UK: xii + 329 pp., illustr.

WORLD HEALTH ORGANIZATION (cited as WHO) (1992). *Women, Health and Development*. (Report of the 1992 Technical Discussion.) World Health Organization, Geneva, Switzerland: 33 pp., typescript.

11

The Attitudes and Involvement of Religions in Population Planning

by

GARRY W. TROMPF, PhD (ANU)
*Professor in the History of Ideas,
Head, School of Studies in Religion,
John Woolley Building,
University of Sydney,
Sydney, New South Wales 2006,
Australia*

INTRODUCTION

Religions, as highly complex phenomena, are constantly creating crises and projecting solutions for the human predicament — often at the same time. Religions have always provided reasons for discord, yet they have also been a powerful source of social unity. They have been available as an impetus and a legitimation even for armed conflict; but they are also great healers of humanity's war-wounds. They are generally very conservative institutionally, although out of them — and partly because of their reactionary nature — have sprung many of the most radical and subversive programmes for the world's betterment. Religions are commonly accused of not being engaged enough in earthly affairs; but when significant events in the human life-cycle occur close to us — especially a death, perhaps a tragic one, perhaps from old age — we sense that religions can grapple with mysteries far more crucial for us than the legislations of governments or the outcomes of economic policies. Now, because the world of religions embraces the issues of life and death so strongly, we quite naturally ask what sorts of attitudes are to be found among them concerning the current population explosion, and what involvement do the adherents of religions have in addressing this enormous crisis?

If, as current estimates have it, Planet Earth may be forced to sustain 11 thousand million humans early in the next millennium, the prospect of very terrible struggles for resources now looms— especially if politicians keep on 'dangling the carrot' of an improved standard of living before the noses of national populations. Some may admit that too many scientific prognostications have made ours the 'most unnecessarily fearful of all generations', while still acknowledging that one glaring exception — 'overpopulation' — should cause us to panic, according to one who has been both a physicist and a science administrator (Farrands, 1993). If so, are religions compounding the problem, or offering an answer; or are they, even while easing the human being's anxieties spiritually, giving us all the more reason to worry over the survival of the human species?

Some strange paradoxes arise simply by posing such questions. Religions have traditionally been the nurturers of group life; could it be that, in praying for health and fecundity, and encouraging love and sexual union between men and women, they are now abetting rather than preventing 'the population bomb' or 'population explosion' (Ehrlich & Ehrlich, 1990)? And could it be that in those instances which most undermine religions' popularity today — in their use of sacred commitments to bolster violent conflict on one hand or, on the other, in imparting certain ascetical modes for repressing sexual desires — we possess, shockingly enough, some necessary (if contrasting) means to reduce population? What conundrums we thus face; yet of course such paradoxes present themselves beyond the domain which one normally identifies to be religious, because the same sorts of questions could be asked of all local, national, and international, agencies concerned with human group behaviour. On one side of things, for example, WHO workers will worry over high mortality-rates in primal villages, or about the lowering sperm-counts among male inhabitants of polluted cities, where the major object of their practical scientific work will be to increase the chances of fertility in a world that is populating too fast! On the other side, one could justifiably contend that a much greater population problem would now present itself if we had not had the many cases of war, revolution, famines, and epidemics, which we have experienced in the twentieth century!

This exemplifies one of the enigmas of our contemporary predicament. At a microscale it is epitomized by medical doctors keeping to their Hippocratic oath of proper conduct; each life which they save only seems to add to a bigger dilemma. At the macrolevel, every disaster which is averted or mitigated by national or international organizations, means that more people remain alive than Nature (or 'normal circumstances') might otherwise allow, as the exponential demographic curve swings for ever upwards. And naturally, because such big ethical issues are entailed here, we would like to know what kinds of responses religions would make to them, and whether their teachings — as powerful influences, being vehicles of collective discipline and restraint — might make any difference.

Let us concede, in advance, that one cannot expect too much from social organisms as traditionalist as most religions are. After all, the major world religions, and also most small-scale local and regional ones, came into existence when not even the slightest prospect of a global population crisis was in sight. Thus all of them in some way or another affirm in their scriptures or lore the value of human increase and prosperity, with large, healthy families being the genuine pride rather than embarrassment of a group. Historically, at any rate, the limitation of populations has not been on their agendas — certainly not as some conscious plan for the comfort of future generations. Only over the last three decades* have members of belief-systems begun to feel international pressures to contemplate overpopulation macroscopically — often to be caught 'off-guard', and not at all in a position of readiness to face up to the effects of drastic world-population growth. There is a difference, here, in respect of environmental concerns generally; for, in so many religions, we discover traditional reverence for their cosmos or natural order, and consequently deep desire to maintain it 'as is'.

With the current threats to the world's basic resources, it is remarkable how many spiritual insights are being found, quoted from age-old wellsprings of wisdom, and propagated to encourage a responsibility towards forests, waterways, and wildlife. This 'greening of religion', as it is now often called, has, however, not usually included confronting population questions within the whole ecological problem. Thus we may note, for example, how little the population issue is raised in the articles in the recent anthology provisionally sub-entitled *The Religions of the World respond to the Environmental Crisis* (Veitch, 1995).

---

* It was in 1966 that we were struck with a kind of vision of the world suffering from far 'too many people' and decided that it was our solemn duty to do all we could henceforth to warn humanity of the foreseeably dangerous consequences both ecologically and otherwise of our species' ever-increasing numbers and profligacy. Although others, such as the Ehrlichs, were being similarly motivated, we abandoned our career (including grants etc.) in explorational research and ecological teaching to found and edit the international quarterly journals *Biological Conservation* (with early hiving-off of *Environmental Pollution*) and subsequently *Environmental Conservation* and generate such other *ad hoc* activities as the World Council For The Biosphere and our Foundation's International Conferences on Environmental Future. — Ed.

The ethics of caring for the environment and ethics involving the reproduction of the human body are more than often treated as quite separate areas of concern. Religion's place in relation to environmental questions is not a simple one, because it can be invoked to improve our control of Nature technologically — not just to preserve wilderness, ecosystems, or wider ecocomplexes. And because religions have been unanimously supporting *life and its increase* in *both* the environment (whether tamed or wild) *and* among humans (at least in their group), the responses of the greater part of their members to the population crisis have been very uneven and widely unconsidered, with only far-seeing individuals or select groups among them carefully measuring up consequences and considering the inevitability of 'limits to growth' (*cf.* Club of Rome, 1975; Polunin, 1980).

Before going further in our consideration of these grave matters, however, a straightforward introduction to the shapes and phenomena of religions will be necessary, for otherwise basic distinctions that are requisite for understanding the various implications of the world's religious scene for 'the population predicament' will be missed.

THE FORMS AND TYPES OF RELIGIOUS LIFE

Mainly for analytical (though also partly for historical) reasons, it is first useful to distinguish the multitude of *'small-scale'*, *'traditional'*, *'primal'*, *'tribal'*, and *'indigenous'*, *religions*, over 7,000 of which still survive into the present (although with most becoming modified in modern times under the 'missionizing' pressures of larger traditions). When investigating the history of religions, significantly, we find that the early attempts to generate 'wider unities' (the empires in Mesopotamia, Egypt, India, China, etc.), were new experiments arising from the building-blocks of small, local traditions that were comparable in both size and their general concerns with the many indigenous lifeways we learn about today (from equatorial Africa, for example, *via* New Guinea, to Amazonia).

Such 'small fry', in the past prejudicially called 'primitive' and 'animistic' religions, have enough features in common for us to treat them as a block, but it also has to be recognized that almost all of them have come under the influence, not only of major faiths, but also of modern nationalisms, and that they are thus *indigenous religions (and cultures) undergoing rapid social change* — a circumstance which we will find to be important to the world population question.

As for the *'great' or 'world' faiths*, at least ten of them — still living religions — can be delimited for the purposes of this study. In

rough order of their historical appearance, they are the Hindu tradition, the Israelite–Jewish tradition, Zoroastrianism, Taoism, Jainism, Buddhism, Confucianism, Christianity, Shinto, and Islam.* Of
these ten, perhaps Jainism and Shinto have only recently, over the
last century-and-a-half, crossed a sufficient number of cultural boundaries to warrant the designation of 'world faith'. And to these ten
we can justly append better-known, *newer world faiths* arising at the
margins of long-accredited religions, and among these Druzism,
Sikhism, and the Baha'i faith, all related to Islam, are the most
distinguishable (although their international profile is also recent).

Matters are not so simple, of course, that one can stop short at
such bare descriptions. For virtually all these *faiths fall into internal
divisions* of one kind or another. It is well enough known that there
are three major expressions of Buddhism, for instance — the
Theravada (especially associated with Sri Lanka and Thailand), the
Mahayana (China and northern south-east Asia), and the Vajrayana
(traditionally Tibetan). More renowned are the major strands of
Judaism (especially Orthodox and Liberal); of Christianity ([Western
or Roman] Catholic, [Eastern or Greek and Russian] Orthodox,
Protestantism, and sectarian Protestantism); and of Islam (the Sun-
nis, the Shi'a, and various sectarian developments). All the world
faiths we have named bear their signs of inner tensions or divergent
emphasis, and any detailed analysis of religious attitudes to the
population question is going to have to take such differences into
account.

Such an analysis, moreover, will only be superficial if it does not
allow for a discrimination between *official and popular expressions
of the major religious traditions*. With the advent of 'International
Relations', religious leaders from the above-mentioned 'world
faiths' have been led to make official statements that impinge on the
population crisis; yet what they enunciate does not necessarily bear a
clear relationship with what is being held and practised among the
mass of their religious followers. Only exceptionally do the 'great
religions' have individuals who are so significant that they might be
said to speak in the name of a whole collectivity: the Pope (the
'Bishop of Rome') for Western Catholicism, the Dalai Lama for
Vajrayana Buddhism, and perhaps the Emperor of Japan for Shinto.

Across the board, however, there are no longer any persons who
can claim authority to distil the views of whole faiths; and even
towards the crucial personages just listed, some contrary stances will

* Note that, to avoid offence arising from this ordering, consensus dates for
Zoroaster (Zarathustra) vary from 1500 to the sixth century BCE; also that Taoism,
although Lao-Tzu is often presented in traditional writings as a contemporary of Confucius (Kung-Fu-Tsu) in the sixth century BCE, is now often thought to reflect a much
older strand of spirituality. BCE = Before Common Era.

be taken by rank-and-file members, or in 'clerical' quarters. At the same time, note should be taken of international organizations (such as the World Council of Churches and the International Network of Engaged Buddhists), or special religious councils and conferences, that produce authoritative views about key issues, or sponsor publications covering them, on behalf of whole masses of followers.

Not all organizations which are internal to any religion, even though they have developed an international constituency, can claim a comprehensive mandate, so that, side by side with formal divisions within religions, we should quickly add *internal divergences of ideological emphasis*, and the factor of significant *vocal minorities*. Competing stresses over belief and praxis are nowhere more obvious today than in the tensions between 'liberal' and 'fundamentalist' approaches to given traditions. The effects on religions of both 'modernization' and liberal–democratic tendencies have meant that, in the presence of modern science, many old dogmas and rituals have been discarded or modified, and 'liberally-minded' elements have welcomed the changes. The availability of The (oral contraceptive) Pill from the late 1950s onwards presents a pertinent example; because the Scriptures of world faiths contain no references to such a mechanism to prevent or delay female ovulation, liberals in all of them accepted The Pill's use by women (certainly within wedlock) as a matter of personal choice.

As a countervailing set of social forces, however, arising partly in reaction against modernism and partly out of a quest for religious renewal, various conservative movements have made their presences felt in the present century — and in defence of old truths and disciplines. We commonly hear many such developments described as 'fundamentalisms', though care should be taken to distinguish one fundamentalism from another. (In Christianity, for example, fundamentalism basically means an acceptance of the inerrancy and literal truth of the Bible, while in Islam it means recovering a situation in which religion [*din*] and politics are no longer separated, so that the Scriptures [*Qur'an* or Koran] can be applied to every sphere of life.)

One current outcry, relevant to population issues, which clearly marks religious conservatism, and to a considerable extent fundamentalism as well, is that against abortion. Mention of the anti-abortionist stand will reinforce the point that, within whole religious traditions, or more often within divisions of such traditions, vocal minorities can mobilize over particular and sensitive causes. It would be unfair, however, to prejudice the positions of whole religious traditions from what is actually verbalized by these 'lobbyists' (*cf.* Hilgers & Horan, 1972).

In these introductory descriptions one should not neglect the many *new religious movements* of the contemporary world. Almost all of these are small in their numbers of adherents, and in that sense comparable with indigenous traditions; yet in the modern world they often represent new sets of ideas that are quite capable of branching out internationally. Many of these might be classed, on close inspection, as minor new sects of one or another major religion; many, however, amount to highly innovative departures, because they *combine traditions* into a new mould. Perhaps they combine 'late-coming' with 'indigenous' strands (as does Subud, for instance, emerging from Java); perhaps they combine two large *traditions* (as Sahaj Yoga does specially with Hinduism and Christianity, for example, originated in India by 'The Mother', Shri Mataji, Guru of Sahaja Yoga); or perhaps they are highly selective and deliberately cultivate the combining of various spiritual strands into a 'New Age blend' (the most complex mix known being found with the American commune 'The Brotherhood of the Sun/Son').

Whatever the impacts of such developments — and in most individual cases they are very slight — together they constitute one of the most significant, if neglected, social revolutions of our time, emphasizing the extraordinary and increasing plurality of religious life. This datum is of obvious importance regarding population matters, because it is clearly going to be harder to pinpoint 'the attitudes and involvement of religions' if the religious scene is losing its old homogeneities and splintering into a less 'analytically manageable' complexity.

Still more relevant phenomena need consideration, and these preliminaries can be completed only with a final, if rather surprising, distinction — between definitive and surrogate religions. The smaller indigenous traditions, the bigger 'world faiths', and virtually all new developments just discussed, can be deemed religions in the normal (or normative) sense, because in them we find world views and activities that are basic to the universally preconceived understanding of what religion is. In them gods, spirits, or a 'Supreme Being' or some transcendent being, will be of vital concern, and with them we are likely to find rituals, special places for 'spiritual contact', along with the teaching of outlooks, deep commitments, and some moral ordering. Yet it would be folly to exclude from view other complexes of ideas and social purposes that can take the place of religion, even though appeals to any deity or a Buddha or distinctly sacred things are lacking; for such appeals can be appropriated away from traditional religious institutions by forces or movements in the 'secular' sphere.

I refer, first, to Communism, which has often been represented as a secular analogue to a 'church', with Marxism as its dogma and final 'arbiter of truth', and the Party possessing the equivalent of an ecclesiastical hierarchy and even a 'Vatican' (the Presidium, etc.).

Our analysis can hardly avoid this 'earthly Faith', the earlier object
of which was to replace religions, and at this point in time more than
one-fifth of the world's population is still under Communist rule
(mainly in The People's Republic of China, the world's most
populous country). Nationalisms (together with types or forms of
civil religion and patriotisms) constitute comparable surrogate
religions, and in fact since Communism failed to break national
boundaries its claims have come to overlap with nationalisms.

With nationalisms, the claims of loyalty to a nation are often
represented as the same as 'divine interests and intentions' (even
through, in reality, the concerns of a state can be made to override
the principles of a religion). Nationalisms are currently highly
assertive. Not only is this noticeable with the 'new world arbiter',
the United States, but from the viewpoint of population studies the
early 1990s have seen quite a few more new countries and would-be
nations clamouring to be heard than in the early 1980s. For some
recent assessments, see Trompf (1993).

With the break-up of the Soviet Empire and European Communism, expressions of ethnic (and pre-Communist) nationalisms
have been particularly manifest, because of desperate struggles over
territory (e.g. as between Armenia and Azerbaijan, and within
Bosnia), and these developments appear somewhat 'neo-tribal' in
impetus. To that extent they can be paralleled to the resurgence of
more distinctly tribal groupings in African states (at the moment
especially in Uganda, Rwanda, Liberia, Angola, Somalia, and also
'Zululand'), to 'race nationalisms' (as epitomized in Sri Lanka), or
to claims to autonomy based on a mixture of reasons, including
religious ones (Basques, Irish, Maronite Lebanese, etc.).

In all highly-committed nationalisms there is some appeal to
sacred destiny or divine right over land, and thus they are special
appropriations of religious sentiments, with their expressions in such
places as those in Africa reflecting the continuing impact of primal
or indigenous religious values within processes of social change. For
a better understanding of social mobilization on religious grounds
around the world, though, one should remember that most of the
'world faiths' have spread among different cultures, and the prior
tribal and ethnic commitments are often quite capable of being revi-
talized — in spite of the influences from cross-cultural faiths.

Other surrogate religions are less political, and often go under the
rubric of 'implicit religion'. On one hand, there are distinctly intellectual positions, such as humanism, and non-theistic or even anti-
religious philosophies, competing with religious faiths (e.g. Comt-
ism as the Religion of Humanity, which still has some clout in
Brazil), and perhaps defensive 'ideologies' for capitalism should be
listed here. With relevance to population questions, protagonists of

Darwinism and Evolutionism (occasionally characterized as a 'creed') need citing , as well as systematic apologists for science as better uncovering the meaning of the cosmos than do religions (a view sometimes caricatured as 'scientism', and surely not without foundation). And we can hardly forget intellectuals in the 'Green Movement', such as the so-called 'Deep Ecologists', who combine science with a reverence for Nature's processes.

On the other hand, there are structures in common thought and life which often look as if they replace religion as people's 'ultimate concerns'. A constant preoccupation with money is one of these; another is 'fanatical' barracking for, and identification with, one or another sporting team or even individual. More generally, and of great relevance to issues of population and development, is what we can call 'cargoism'. This is a social reality suggested by the celebrated Melanesian 'cargo cults', in which indigenous peoples call upon their ancestors or deities to bring in the new commodities, meaning European-style goods. Cargoism denotes the growing sense that families have a right to a certain range of the newfangled internationally-marketed goods, that range being specified by local expectations (for previously Stone Age peoples, metal utensils are a top priority; in the West virtually every family hopes for a refrigerator).

Cargoism involves a family's or group's projections of 'comfort' and 'security', and of an improved social status when one's access to the newly-desired commodities is possible. Some of the new items also seem to possess a religion-like power in themselves (note the possibilities of 'inspiration' from electronically-enhanced music or from television). Some analysts claim that control over the new (largely post-World War II) technology means more control over minds than the old religions could ever have had. It is thus fitting to conclude these comments with this treatment of 'cargoism', for, although it can only be deemed a kind of substitute religion or popular ethos, it has quickly become today's practically world-wide religious-looking activity — with considerable implications for the population crisis (Trompf, 1990 pp. 10–4).

THE RESPONSES OF RELIGIONS TO POPULATION ISSUES

This section surveys attitudinal responses of the different religions just introduced to the present-day population explosion.

Indigenous Religions' Responses for Increase

Separate, small-scale traditions, unless they have stolen the media limelight (in the manner of the Hopi Indians in the United States), do not normally find themselves in a position to present programmatic solutions to global problems — although clusters of indigenous custodians, and scholars seeking to publicize 'the wisdom of the [tribal] elders', have tried television networking to propagate the insights of small-scale cultures on ecological problems. Regarding population, what is most impressively and frequently conveyed by these traditions is a sense of biocosmic unity. Primal spiritualities assume an intimate relationship between human and other animal life, and have a keen sense of the interdependence of the components in their immediately-involved environment. A group's survival is, typically and most reasonably, recognized to depend on a balance being struck between human needs and accessible resources.

Two relevant, yet apparently opposing, configurations are manifest. On one hand, such tribal or small-scale group orientations are quite naturally survivalist. A plethora of rites among them seeks the fecundity of women and the avoidance of shameful barrenness, the fertility of the ground, or the increase of game and/or herds for the group's health, as well as the strengthening and protection of men in 'warrior societies'. One simply has to get used to the fact that the expression of these basic desires is endemic to humanity, and has been reinforced or adapted in the great traditions. These hopes are partly enshrined in the UN Declaration of Human Rights — to be fed, clothed, and granted personal security — and they are actually the essential stuff in which theories of Development are framed.

Survivalism and the ritual enhancement of fertility, however, are also principles working towards strong population growth. A prayer among the West African Akan asks, for example, that those about to marry should have 30 children! Although this may be hyperbole, yet special Akan celebrations are arranged for women who bear a tenth child, or twins. This illustration reflects the broader black African understanding, furthermore, that there is no greater social scar than sterility (or the production of a genealogical dead-end), with even spontaneous (let alone induced) abortion being abhorrent (Sarpong, 1977 pp. 58–9). The case also reflects a world-wide preoccupation of indigenous religions with population increase.

Responses for Population Control

On the other hand, such preoccupations betoken the harsh realities of long-inured, even 'stone-age', conditions of subsistence. Not only high infant-mortality rates, but also the hazards of war and sorcery, really make religious accentuations of fertility more balanced than they might at first seem. Moreover, among indigenous peoples a huge body of details beckons examination as to how group populations were kept within bounds. If the Akan immediately killed defective newborns, for instance, the Yoruba (also of West Africa) traditionally destroyed twins. If infanticide was known — for survival (among some Australian Aboriginal groups, while moving), or for ritual purposes (Roviana, Solomon Islands), or to bury beside a killed parent (Auca, Amazonia), or as a matter of parental choice (Ikung, Africa) — so also were there procedures for the less desirable or less productive persons to depart from the 'survivalist circle', such as common trouble-makers being put in the front line of battle (Wain, New Guinea, for example), or aged persons wandering off into the wilderness (as with the Inuit [Eskimo] into the arctic cold).

Plant abortifacients were widely known (e.g. in Micronesia), while taboos, such as those preventing sexual intercourse with a lactating woman, kept children spaced (Gimi men in highland New Guinea, to illustrate, being ashamed at the birth of a second child before the first one is weaned). Moreover some peoples only allowed men to marry late in life (Kiriaka, Bougainville), while others prevented children from being born out of wedlock by sewing up young women's labia until marriage (Meroitic groups in the Sudan). Many of these procedures may seem uninspiring, and, along with a range of unhealthy or violent-looking practices, they could lend seeming legitimacy to missionary activity by world faiths; yet indigenous religions offer the richest variety of responses to population growth, and any international attempt to meet the current population crisis ought to show 'anthropological responsibility' towards such cultural divergences.

The symbiotic interconnectedness between primal peoples and their environments have led some researchers to consider human generation, subsistence, and morbidity, as parts of an ecosystem or wider ecocomplex in which they are commonly linked to cycles of animal herds, the seasons, and natural fluctuations in the availability of usable resources (e.g. Vayda, 1976). So far this approach has been relatively unsuccessful, for it fails to account for the changing efforts of human purposefulness; but it does rightly intimate the greater, almost inevitable, attachment of each indigenous people to their 'cosmos', which is, of course, a highly confined one. Such localized awareness, however, undergoes serious modification under the new internationalizing pressures to which primitive peoples are liable nowadays to be more and more subjected.

Indigenous Religions Undergoing Social Change

As soon as 'biocosmic' rhythms are disturbed by new technologies and horizons, many of the highly culture-specific customs and practical responses start to lose their hold and some become highly vulnerable to the charges of being mere superstitions and functionally inhumane. On the other hand some of the old, ingrained general attitudes remain — such as those concerning fertility and security — and while these are to some extent reinforced by introduced religions, which (as we have already stated) are also presented as *for* Life, new pressures and fixtures often have to be faced, including the introduced dangers of colonial peoples' diseases and newly-established institutions — such as hospitals, family-advice clinics, etc.

In some cases populations increase very markedly because of such changes. Thus in the New Guinea highlands, after World War II, warrior peoples were substantially pacified so that the numbers killed in inter-tribal war or raiding were reduced, and males had more time to participate in family life (as encouraged by missions), because their manual work was greatly reduced by the new technology (steel axes rather than stone ones, for a start, being used to make garden fences). In the meantime, Western medicine removed the curse of yaws, wound ulcers, and (where they existed) malaria and the *kuru* disease, while the same medical system also countered the 'whiteman's sicknesses'. With age-old preconceptions about fecundity still applying, it was not long before Papua New Guinea held the fastest-growing population in the world (1975–7) (Trompf, 1994 *esp.* pp. 433–7). If, in this case, the high tropical valleys were able to sustain this new growth, and sufficient clinics were established to lessen infant mortality and death in childbirth, the same cannot be said of other countries, where declines of mortality produce new environmental strains, and where lack of facilities means falling back towards the earlier 'low standards' (WHO, 1984).

Where and when chances of survival are *decreased*, the spiritual and psychological pressures are to multiply, so that some children will survive; these are the typical responses of indigenous world-views. Such outlooks and outcomes tend to persist when social change — including civil and international wars, or such an awesome epidemic as AIDS — bring, or threaten to bring, new forms of hardship. Problems further consequent on trying to multiply in circumstances that are already arduous, moreover, are usually not seen as the responsibility of the strugglers themselves. In character with indigenous outlooks, these problems are blamed on 'outside influences' — locally on the work of sorcerers, regionally on other tribes, and nationally on neighbouring nations or external 'imperialists'. It is extremely difficult to persuade indigenous people in dire socio-economic straits of their *own* folly in excessively reproducing. Outside observers quickly understand and sympathize with the indigenes' position, and aid workers also tend to shift the onus away from individual choices to matters of structural violence, while helping to keep alive those who require their assistance.

In certain kinds of hardship, however, indigenous peoples are also ripe for religious change — Untouchables and tribals in India, for instance (for there are around 400 small-scale 'traditions' in India alone), have looked to Christianity in this century as the route out of the caste-system; and Latin American blacks have recently made a massive turn away from Catholicism and more traditionalist spirit cults to sectarian Protestantism to get them out of the 'poverty trap', while in the urban north-east of the USA many 'African Americans' have been attracted to Black Islam for the same reason. It behoves us now to look at the greater religious traditions and their spread, and to gauge their implications for global population growth.

'OFFICIAL' STATEMENTS OF WORLD FAITHS

In accordance with the mode of analysis suggested earlier, official and popular beliefs and practices will be considered separately. The only publicized attempt to collate 'official responses' of world faiths with the population issue was that made in 1990 through the World Wildlife Network on Conservation and Religion (WWF, 1990 pp. 2–3). It failed to be truly representative but nevertheless was an important beginning. In what follows, quotations from these statements will be referred to under the abbreviation WWF, and an effort will be made to rectify imbalances and explain both the status and important intricacies of different positions. Each tradition will be considered in the order of its historical appearance (*see* above), except where slight adjustments are necessary to allow for improved analytical flow and for some allied cases to be taken together.

The Hindu Tradition

It is false to speak simply of Hinduism (over 500 millions strong) because the religious groupings derived from the ancient Vedic tradition are highly diverse, consisting of some large 'schools of practice' (Vaishnavites, and Shaivites) or 'thought' (Vedantic), and a vast number of minor ones — some highly ascetical, some very fertility- and even sexually-focused. The most popular Scriptures, though, set the main tone by extolling the procreative powers of beneficient gods or heroes. Although these model beings were more active or incarnated in superior Ages or eons, *i.e.* before the present one of evil and adversities, they offer ideal images of big and status-producing families. Sons are important for carrying out ritual functions after their fathers' death, while the female reproductive power is sanctified by Brahmanical (and thus priestly) regulations against abortion (*Laws of Manu* **5**: 90; **11**: 88).

India, where the vast majority of Hindus are located, is currently the world's second most populous nation (850 millions), but the problem of overpopulation, and of the mass poverty connected to it, were not so clear until the late 1920s, when the population was still very much smaller than it is today. Since then, Hindu religious leadership has not reacted in a uniform way. Some factions have acquiesced to the post-War independent government's programme of family planning and sterilization (operations that have for some time looked less successful than formerly) (Srinivas & Ramaswamy, 1977); others have opposed contraception as being contrary to the principle of non-violence towards life (*ahimsā*). Appealing to *ahimsā*, some spiritual writers maintain a defence for the overpopulous sacred Indian cow as the animal best alleviating — through milk and butter — problems attendant on human increase (*e.g.* most recently Saraswati, 1993; *cf.* Chapple, 1993). Other leaders again keep locating the cause of the population problem in wrong orientations of life, teaching against materialism and its illusions (*māyā*), and trying to revitalize old virtues of asceticism and the celibate 'masters' (*brahmacāryas*). Assessing the global situation, a Hindu spokesman, Ranchor Prime (WWF, 1990), recently turned the heat on the wastefulness of the West, maintaining 'that overpopulation is itself not a threat', so much as 'the greed of modern industrial society'.

Judaism (the Israelite–Jewish Tradition)

The imperative to 'be fruitful' is writ large in the Hebrew Bible (= the Christians' Old Testament), especially in the Book of Genesis (esp. **1**: 28–9; **9**: 7). As a result, it is a Jewish duty to marry — the very mark of being fully a man and a woman, in fact — and to have children. As we saw with the Africans, childlessness in a Jewish family is considered a tragedy. The ancient promise to Abraham of a truly multitudinous nation (*e.g.* Gen. **22**: 17) always hovers in the background of Jewish family and social life, and the will to be populous as God's special people has been very much affected by two crucial developments in the twentieth century, one terrible — the 1939–45 'Holocaust', or Nazi programmes of extermination — and the other more promising, but chronically disruptive as things turned out, namely the creation of the modern state of Israel (from 1948). In view of the former tragedy, Jewish leaders' statements about their population numbers (approx.14 millions) have been much more about survivalism than about their people's over-population. In Israel itself the Ben Gurion Prize was awarded to women who bore a tenth child, at least until the early 1960s, and in the 1970s even state rhetoric, neutrally referring to 'Families with

Multiple Children', changed to 'Families Blessed with Children' (women receiving additional state benefits for each new child), cf. McDowall (1993 p. 648). Today the pressure for Israeli Jews derives more from public recognition that the numbers of Arabs in and on the edges of Israel are increasing faster than they themselves are (cf. Yuval-Davis, 1989 p. 96).

Particular Jewish regulations — what can be called 'orthopraxy', or the right practice, which is especially important for Orthodox Jewry — are, on balance, leaning more towards 'healthy increase' than 'restrictive decrease'. Along with non-marriage, celibacy, monasticism, homosexuality, masturbation, *coitus interruptus*, as well as abortion, contraception, and artificial birth-control methods, are all considered reproachful in tradition. But some of the older regulations about abstinence tend in the opposite direction — such as refraining from intercourse for five months after a birth, or during some ritual periods — and debates do continue about all the previously-listed issues. Pronouncements in the *Talmud* (rabbinical commentaries on Jewish law) have it, for example, that a foetus under three months in development is 'just water' rather than human, abortion of it being permitted if birth would be seriously disadvantageous to the mother — especially to her health. In modern Israel, rabbis acquiesce to government sponsorship of Family Planning Clinics, and to women's access to Committees that officially permit abortion, though these institutions are not tolerated in religiously sensitive centres, such as Jerusalem (with the threat to remove *kosher* food from hospitals, etc., being used against those trying to introduce such secularities).

The effects of expanding populations on environments, living standards, and the global future, however, have just begun to generate a new area for debate. As Aubrey Rose, Chairman of the Jewish Working Group on the Environment in the UK, put it, Jews will have to balance the old injunctions to multiply with their 'environmental precept, "Thou Shalt Not Destroy"', if 'a population explosion' would lead to 'the destruction of nature and human dignity' (WWF, 1990).

Zoroastrianism

Of all the world faiths, Zoroastrianism is the only one faced with steadily declining numbers (now down to 80,000), and there is constant talk of survivalism — particularly among pockets of refugees from Iran who are scattered around the Western world (Hinnells, 1984 p. 48). Their demographic crisis owes itself to suppressions in Iran, the disallowance of Zoroastrians to proselytize in both Iran and India, and the removal of membership and its rights

from the families of daughters who marry out of the Zoroastrian
tradition (a rule that has now been relaxed). This is also proportionally more a religion of the well-to-do (even in India, where the
Zoroastrian Parsees are economic leaders in and around Bombay).
Traditionally, in any case, three children per family were considered
enough by Zoroastrians, and after that women 'had to be careful'.

The same formulaic prayers which the barren would offer to God
(*Ahura Mazda*) for a first child were offered by mothers to avoid a
fourth. This tradition is now known more for its emphasis on
permissions of ethical choice than on strict moral regulations (*cf.
Gathas*, Yas. 5), and using birth-control techniques was a personal
matter of seeking 'right actions'. Women pressing a special case for
an abortion have been respected, traditionally being fed a mixture of
ground seeds and herbs, or sometimes 'shocked' into miscarriage
(*e.g.* by concocting bad news). The self-limiting aspects of this
tradition are its own advertisement as a solution to overpopulation.

Jainism and Buddhism

These two traditions, as ascetical movements breaking away
from the Vedic–Hindu tradition (and thus traditionally considered
'heretical' by the Brahmin priests), are both inevitably 'passive'
regarding efforts to control fertility. Their teachings of compassion
towards 'all sentient beings' (as Buddhists put it), even 'all life-
forms' (thus the Jains), require avoidance of any action (*karma*) that
takes away life (so abortion, and certainly infanticide, are generally
opposed). On the other hand, some nerve-centres of leadership in
these traditions have become quite proactive over environmental
issues of late (especially through the radical 'Engaged Buddhist'
network), and 'religious solutions' to the population dilemma are
now being more confidently stated than heretofore.

With the Jains, so consistently a small population (largely in
India), it is easier for leadership (both monastic and lay) to be
representative, and also easier for it to defend this 'self-limiting'
tradition as a solution! Self-restraint and *ahimsā* are so essential to
Jainism, in fact, that the high ideal for life's completion (*i.e.* as one
gets older) is to subsist without food and drink (and thus any
harmful effect) unto death. 'Population control' is attendant upon
'self-control and transcendence of passions', which is expressed by
the monks *par excellence* (two major orders), but also in the moderate and monogamous conjugalities of lay couples (quoted in WWF,
1990). But note that, just as we would have expected the strongest
statement of population control from this remarkable grouping,
special globe-trotting monks are now making contact with scattered
Jain communities and beginning to celebrate the growth of their
membership (to over 8 millions)!

Buddhist schools, for their part, abound in traditions of self-abnegation and the renunciation of earthly love for the sake of Enlightenment (as the Buddha himself, once a pleasure-loving prince, exemplified). Sexual desires, then, or the want to make one's mark with a large family, can very easily be recognized as forms of craving (*tanha*), which cause the very 'suffering' or 'unsatisfactoriness' that the famous Buddhist *Fourfold Path* is set to eliminate. An increasingly world-wide voicing of Buddhist principles (the speeches of the highly mobile Dalai Lama among them, and with 300–500 millions concentrated mainly across ten Asian countries expected to heed them) suggests that all social problems, including the population one, can be solved through the frugality and simplicity of life that constitute what Buddhists call 'right livelihood' (Eppsteiner, 1988). Through social pressure, and even political policy (as in Thailand's case), youths are encouraged to experience 'monkhood' as novices, this celibate role having obvious implications for limiting population.

Taoism and Confucianism

With its followers scattered through eastern Asia, Taoism is a rather diffuse body of belief, and is highly decentralized (and thus generally lacking in official representation internationally). This tradition shares, in common with Jain and Buddhist views, a sense of cosmic processes whereby people's spiritual maladjustments will bring on 'their own problems'. If, in the two Indian-originated traditions of Jainism and Buddhism, the relevant discourse often focuses on reincarnation, or on the incessant wheel of life (*samsara*) which most individuals get caught up on through performing varying amounts of bad actions (or *karma*), Taoists speak more of the unfortunate consequences of disequilibrium. All personal and social problems result from the creation of some imbalance (*cf. Tao Te Ching* 42): even too much good can bring evil against it. Thus overpopulations, in the manner of epidemics or even earthquakes, are calamities that come because of failure to attend to matters of cosmic harmony. In the contemporary situation, this is an implicit policy of wise restraint.

Confucianism, though formerly more politically powerful in old China than elsewhere, contains no clear guidelines to address overpopulation (which has only been a modern problem for eastern Asia). Its scriptural classics contain nothing specifically about birth-control, even though the human family becomes the model in Confucius's philosophy for the ideal ordering of states (a position sometimes called 'Familyism') (Kitagawa, 1960 pp. 86–98). If anything, what is stressed as the virtue of 'filial piety' puts a priority on having sons to succeed as family heads — a principle not lending itself to a consistent limiting of births. The medicinal–herbal lore for which China is renowned is more concerned with fertility over-coming barrenness, and securing a male progeny, than with pre-venting more births or causing abortions, although medicines did exist for these latter purposes and using them has become in-creasingly a matter of choice. In modern-day Taiwan, mainland China, and elsewhere, Confucians (approx. 350 millions) acquiesce to the pressure of family-planning clinics.

Shinto

Although Shinto (with over 100 million members) possesses its socio-political centre in the Japanese emperorship, and although there is a body of common customs reflecting its impact, as a religion its beliefs are unsystematized and its practices are often localized to temples and their domains, which have been patronized by leading families for generations. In Shinto there are a number of background cultural factors and attitudes with implications for population matters. The glorification of the male hero is probably stronger in Japan than anywhere else in the world, and tends to foster family hopes for the birth of sons. Also, because of the long-standing customary methods of reducing child numbers through abortion and, earlier, infanticide, Shinto — among all the religions in Japan — is the tradition most lending itself to the high abortion-rate there (overall the highest in the developed world, with the 1948 Eugenics Protection Law making abortion legal 'under rather broad conditions'). For background see Lorimer (1954 p. 234). It is striking to see how adherents to other persuasions in Japan, includ-ing many Buddhists, have bent with the cultural tendencies, even though the Japanese are noted for conformity to prevalent attitudes or rules imposed from the top. The 1992 Health Ministry Report (ISEI, 1993 pp. 79–81) expresses concern at how the low birth-rate in Japan may produce 'different' trends in reaction.

Christianity

Christianity (more than 1,000 million adherents) is, as the largest of the world faiths, deeply divided over aspects of the global population predicament, and for that reason there is more lively controversy, theological positioning, and specialist literature, about the subject in Christian, than in any other, circles. The years 1959–60 saw the first serious monographs about overpopulation emanating from among any of the religions, with the Catholic French (Jesuit) Stanilas de Lestapis's *La Limitation des Naissances* (1959) and the Protestant (American Congregationalist) Richard

Fagley's *The Population Explosion and Christian Responsibility* (1960), while the latter was working for the World Council of Churches. The official divide between the Western communions (respectively Catholic and Protestant) over the birth-control question sharpened with the publication of the Papal Encyclical *Humanae Vitae* in 1968 (Paul VI) — which declared against artificial methods of contraception (and reinforced Pope Pius XI against abortion) — and the clarification by the 1960s that most Protestant churches accepted the adult individual's personal right to contraception and saw its importance for checking overpopulation (*cf.* Fischer, 1965; EKD, 1994, some denominations even showing more lenience towards social as against medical grounds for abortion).

Catholic teaching is very much affected by the Old Testament imperative to be fruitful and to guard life, and Christ's re-validation of the Commandment against murder (Mark, **10**: 19) militates against liberal policy over abortion. Official mainline Protestant positions have stressed more the response of Christian love to whole contexts, admitting certain 'valid' social circumstances which 'justified birth-control' (especially extreme poverty, *cf. e.g.* 1988 Lambeth Conference, quoted in WWF, 1990; *cf.* Draper, 1972, pp. 204–8 for earlier Anglican Statements). Orthodox leadership, amid discussions of theological opinion, leans more towards the official Western Catholic line, while most sectarian Protestant groups (*e.g.* most Baptist groups) accept personal rights over contraceptive methods but loudly inveigh against abortion.

Certain technicalities make it difficult to fathom the current crises over birth-control within the Roman Catholic fold. To understand it better, it must be remembered that no Papal pronouncement on this matter has so far to be taken as infallible or an article of faith (*de fide*), because none has been made in or through a Church Council or in collusion with the majority of bishops.* This means — as various Catholic ecclesiastics who are worried about the currently-stated position have been quick to argue — that the Catholic laity are, thus far, left to explore their own consciences on birth-control. On the other hand, conservatives have been only too eager to retort that the statements of Catholic leadership over the centuries has been so unanimous on the subject that the Papal Encyclicals instructing on birth-control — including John Paul II's 1993 *Veritatis Splendor* — should be taken as wellnigh mandatory for the faithful (for background, *see e.g.* Sulloway, 1959; Küng, 1972 pp. 31–55).

---

* This gives us hope for near-future improvement in this vital matter where practising Roman Catholics are concerned. — Ed.

Following effective campaigning in a number of countries —
witness Anne Biezanek in Liverpool, England, UK, during
the1960s; Biezanek, 1964),), Catholic family advice services have
become more frequent. Practical advice sets much store in the
rhythm method as 'scientific' (*i.e.* no likely conception if sexual
intercourse is confined to those 'safe periods' before and after
female ovulation), and thus also stresses the discipline of self-
restraint (Noonan, 1986). Apologists for the official stand often
maintain that Western Catholicism (the largest-in-numbers single
component of Christianity) is responsible for a lessened rate of population increase than applies to other religions (which appears true
when the comparison is with India, but stands as a problematic claim
with regard to Latin America and the Philippines). Church working-
parties and research units from time to time address various ethical
issues relevant to birth-control — including euthanasia, *in vitro*
fertilization, genetic 'engineering', and the effects of the 'permissive
society' (especially on young people) — most of these debates being
organized by the Catholic community or member churches of the
World Council of Churches, but some also by the Society of Friends
(Quakers) and Unitarians, though they have less financial resources.

Islam

As the second-largest religious tradition (approximately 1,000
millions), Islam also contains inner tensions in its reaction to threats
of overpopulation, whether regional or global. By the late 1950s a
number of family-planning clinics were accepted to meet crises in
Egypt and Pakistan (though simultaneously rejected in the more
secular but far less populous Turkey). In 1953 the Fatwa Committee
in Cairo, authoritative on Sunni canon law, declared that the use of
medicine 'to prevent pregnancy temporarily' was not debarred (the
Qur'an [Koran] not requiring hardship from believers, *Sur.* 2:185;
22:78), even if medicine 'to prevent pregnancy absolutely and
permanently' (= sterilization) was forbidden.

Other arguments from the Scripture and tradition of Islam are for
granting the right of birth-control to feeble women, poor house-
holders, or even women who are fearful of losing their beauty
(Lestapis, 1959 pp. 42–7). In pre-revolutionary Iran (under the Shah),
special education teams went out to rural areas to advertise new
options, and at present there is a slight shift back in that direction —
with large billboards, for instance, asking passers-by 'Do you have
time to spend with a lot of children?' and 'Can you give more love to
[fewer] children?' In Indonesia, the nation with the largest Muslim
population, the Islamic councils have openly (if cautiously) lent
support to government family-planning programmes (*cf.* Iskander,

1974), as they also have, but without much effect, in Pakistan, (Akbar, 1994)..

On the other side of the coin, there are currently ideological pushes for strong, increased Muslim populations, more politicizing (*i.e.* fundamentalist) Islamic leadership advocating world-wide expansion in population terms (and sometimes in antipathy to the West). This is often coupled with an awareness that Muslim nations are disadvantaged in comparison with the West in terms of high mortality and morbidity rates (for relevant insights *cf.* Bah, 1992), and at times with a recognition that super-rich Muslims are not meeting the Qur'anic (Koranic) challenge to care for the needy. Thus we find, anomalously enough, that in a country of such unequals as Jordan, most family-planning centres turn out to be facilities where urbanized Muslim parents are helped to have babies, certainly not to prevent births, and yet there are significantly impoverished (and desert) areas of the country where deaths in childbirth are high (for lack of medical institutions or adequate nutrition).

In contemporary Iran the enunciations of the late Ayatollah Khomeini still resound that 'Those who say 'God is Great' [*i.e.* committed Muslims] should be more and more, day by day'. This was a challenge to the general masses (Islam being presented as 'a religion of the humble, not the rich'), and Iran's population rose 3.7–4% yearly during the Revolution and the Iraq–Iran War. In both Iran and Iraq, of course, encouragement of government and clerical pressures still remain to make up for population losses during that war.

Other World Faiths

Among the other, remaining traditions, Druzism, Sikhism, and the Baha'i faith, have variant approaches to population growth. Decimated in the past, and with only about one million adherents, the Druzes are mainly survivalists at present, happy that their prosperous families outside the Middle East are enjoying larger-than-average families (4–5 children). The Sikhs, for their part, are seriously divided in southern Asia itself, but they have usually rejected 'intrusive and coercive methods of family-planning' (tried in India), and advocated personal restraint and the heeding of '*His Hukam*' (the divine will) if the human race is not to be out of balance with Nature and 'suffer the consequences of its own actions' (according to the UK Sikh Education Council, *see* WWF, 1990). The Baha'is find the solution in adoption of the Universal values advocated by their prophet Baha'ullah: through establishing a world community or government based on these standards, the implicit promise in their Scriptures of a highly-advanced civilization would be achieved.

POPULAR AND UNOFFICIAL ATTITUDES OF WORLD FAITHS

Non- or less-official approaches to overpopulation, that announce more restraining positions than those found in the major positions just discussed, are very much in the minority — although ascetical or monastic alternatives, or schemes of decisive social engineering framed by isolated intellectuals and visionaries, do have their attractions, and are often loudly proclaimed in print or by the aural and visual media as challenging the world's drift towards major crises. In the main, however, populaces under each of the 'religious umbrellas' denoted above tend to think and behave less rigorously than the leadership in their traditions ostensibly require of them. Researchers need to be aware of whole bodies of popular attitudes that religious leaders usually seek to modify or combat — even though, it should be noted, these attitudes are often tangled up in 'worlds of official belief' themselves.

To unravel all the complexities here is impossible in the space available, yet a few important pointers will be valuable. First, one can talk of the *drag of primal fertility interests on major religions*, especially in the so-called Two-thirds World. In simple terms, this means that, in certain regions where world religions have grafted themselves on to tribal traditions, the latter have tended to appropriate the former as reinforcements of their own earlier concerns about 'increase'. Thus, to illustrate, the Buddhist saint Upagupta has been transformed into a quasi-deity for bringing in fertility throughout most of northern south-east Asia; while in some southern European and Latin American settings the Virgin Mary seems to be 'culturally translated' into a Mother Goddess (again with connotations of fecundity). The 'pundits' among Buddhists and Christians will have to take this lag into account as a challenge; meanwhile analysts of religious attitudes will have to appreciate the force of these 'grassroots' (*i.e.* bottom-up) realities. Especially in Third World contexts, these underground attitudinal pressures also include *fatalistic notions*, such as the resignation towards every pregnancy or birth as happening always through 'divine will' (Montefiore, 1970 p. 153). Other collective social statements in contexts of disadvantage are those stances which are *blatantly reactive against imposed measures of restraint* — responses that require careful watching.

Thus the poor, perhaps more through radicalizing elements than through arriving at a common understanding of their own rights, react against sterilization and birth-control programmes that appear to be forced on them and not on the rich. As a woman running a Philippines hostel put it, against the policies of the Aquino government:

'Social justice ought to bring about peace, prosperity, and love... sharing and trust in God and in one another. Social justice means jobs, land reform, and economic services. Too often we settle for condoms, IUDs, and sterilization' (Verzosa, 1989 p. 66).

In a recent conference, indigenous intellectuals were arguing the thesis that the West (and in particular the USA) is 'pushing birth control on Third World countries to murder the unborn and stifle the population growth' of tribal peoples and poor nations (Cribbs, 1992). Rhetoric in Islamic countries, that birth-control measures are Western introductions moves in the same direction (and this is the very [socially popular] reason why these measures, along with leniences over abortion, were outlawed during the Iranian Revolution). At this difficult juncture (including rights claimed for the unborn) apparently appeals to rights can *work in an opposite direction from efforts at solving Planet Earth's problems resulting from overpopulation.*

Another unofficial pressure related to social disadvantages is *socio-religious competition for numbers.* This can occur *within* a tradition: the less well-off Sephardi (non-northern European-originated) Jews in Israel, to illustrate, see that they can improve their political clout by population growth (Sephardi family sizes typically doubling Ashkenazi [northern European] ones). Or else the numbers' competition can amount to an *interreligious* struggle: sporadic local crusading among Muslims often puts the case in Bangladesh (of all countries!) that Muslims should have bigger families if they are not going to be swamped by the ever-multiplying Hindu 'idolators' in the long run.

Across the whole world of religions, *millenarian* and *eschatological expectations* also have to be accounted for as attitudes towards the population crisis. In the Jewish–Zoroastrian–Christian–Islamic axis there are shared popular hopes among the rank-and-file that humanity will face cataclysms and cosmic struggle between good and evil forces before God's final interaction (and transformation of the world for the better). In Hindu–Jain–Buddhist thought, the notion of cosmic betterment after this present Age is also constantly percolating, and sometimes (as also happens in Taoism) takes on sharpened millenarian-looking overtones. Such thought implies the attitude that solving global problems is beyond humanity's powers, although paradoxically enough the 'millenarian temperament' has sometimes been a power-house for radical change. Spiritual leaders looking to well-planned practical solutions, however, based on 'solid' research, will usually see in 'fundamentalisms that assume the imminent demise of the earth' a dangerous irresponsibility (*see e.g.* the strictures against such a danger in Osborn, 1993 p. 27 *et passim*).

In the so-called 'First World', popular attitudes of resignation sit alongside (and mix with) those of collective responsibility. Christians in the Post-war West have lived with chronic reminders of mass starvation in the Third World, and while most responses to other people's catastrophes have been those of generosity, minority views would have it that such developments are inevitable purgations of the Earth under strain, or are lessons that other peoples must learn without false hopes being let in by 'band-aid' solutions; or perhaps they may even be 'divine retributions'.

Many conservative or fundamentalist Christians hold AIDS to be divine judgement against rampant sexuality, though they will then divide in applying this approach to the complex African situation (even if various African Christian groups also make sense of this pandemic in terms of God's chastisements against evil). On the side of active responsibility, however, Jewish and Christian prophetic challenges to the well-to-do West have not produced (and perhaps in all reality cannot produce) such a massive shift of socio-economic and political responsibility that inequalities exacerbating the population problem will be seriously eased. Bettering others' standards of living has for long been widely publicized as the means to lower infant mortality and thus constrain survivalist tendencies as it would bring family sizes down; but the West does not yet see that any tightening of its own belt will be a future necessity.

In the industrialized First World, increased educated awareness concerning the population question has also brought *popular sentiment against Catholic inflexibility* over the birth-control issue. Even within Roman Catholic circles, especially in the USA, there has been a significant enough 'rebellion' against the ban on contraceptives to lessen the old statistical differentials between Catholics and non-Catholics over birth-control methods (*cf.* Bouvier & Rao, 1975). On the other hand, one of the most vocal minorities at the moment, both within the USA and beyond, and expanding across 'denominational boundaries', is the Right to Life Movement, which is implacably opposed to abortion (except in extreme medical emergencies), let alone its legalization. Thus wider acceptance of the view that takes ecclesiastical expectations of restraint to be too unrealistic, is at the same time being met by rigorous stances against taking life when once conception has occurred.

NEW RELIGIOUS MOVEMENTS

Although most new religious movements are small *vis-à-vis* the large traditions, collectively they can be said to accentuate the population problem. They are typically survivalist, which means that they want to add to their numbers as quickly as possible to make their mark (although this could be done by missionizing more effectively than by breeding). They are also commonly millenarist, projecting cosmic catastrophes before the coming of the Millennium (allegedly bringing retribution to 'the evil ones' of this world). Ironically it was Rajneesh, *guru* of the Sannyasins, who first predicted an 'AIDS apocalypse' (Palmer, 1986 pp. 8–9), though himself allegedly dying of the disease). Shri Mataji (The Mother), also, as spiritual leader of the Sahaja Yogis, has apparently predicted that half the world's population will be decimated through the spread of AIDS (an estimated one million people now being infected in India alone, and several millions elsewhere), and her followers often explain global overpopulation as the desire of so many souls to be reincarnated as humans while the Holy Mother is on the Earth.

Other new religious movements compete for numbers when seeing themselves within a tradition — as do the Lubavitch Jews, known for having families between 15- and 21-strong. Yet others combine remarkable growth with socially conservative alignments — as in the case of the 16-millions-strong neo-Buddhist Soka Gakkai movement in Japan, which advocates a state religion (for background *see* Ikado, 1968 pp. 106–8, 114–5).

Overall, the effects of the development of new religions on world population will be either highly localized or spread too thinly internationally to be of significance, while the attitudes purveyed through them also produce a very 'mixed bag' (*e.g.* Jackson & Teague, 1978). Those linked with New Age sensibilities are the ones which seem most likely both to stress, and act out in their lives, a balance between human numbers and the environment. Great hopes once prevailed for the Commune Movement; however, as an alternative strategy for Western living, it has been offset by the 'drug culture'. Indian communalities (*ashrams*), as images of environmental responsibility, remain perhaps the most inspiring. However, it is scarcely possible to foresee any of these as at all widely 'saving the world'.

SURROGATE RELIGIONS AND QUASI-RELIGIONS

Special comments should be made about Communism and various nationalisms, and then some general remarks about other attitudinal trajectories. Marxian *Communism* was framed on the premise that scientific socialism would solve 'the Malthusian problem'. Thomas R. Malthus, a late-eighteenth-century Anglican clergyman and Cambridge don, was worried that human populations would outrun available resources: Karl Marx and Friedrich Engels always maintained that such a prospect could be averted through the collapse of capitalism and the advent of an egalitarian, eventually classless,

society. In a semi-religious vein, Engels ([1878–83] 1954 p. 308)
argued that Nature would take retribution on those who, infected by
the capitalist mentality, sought to 'conquer the environment'.
Whether Second World (or Communist) nations have been true to
the Marxist vision or not, they have typically taken strong state
action to affect their population situations in one way or the other,
and have claimed legality of their approaches as 'orthodox' ones.

With the breaking up of the USSR, comments here will be
confined to contemporary communisms. Cuba has encouraged population growth, pointedly contrasting its healthy growth-rates with
those of its non-communist Latin American neighbours. China's
current predicament reflects certain contradictory moves: a longer-
standing concern, from 1956, to reduce population growth, and a
'hiccup' in the contrary direction in 1960–1, when Chairman Mao
was proclaiming that China would develop more if its people
multiplied more, until the return to 'social engineering' in 1962.
Since this last date, provinces have framed rules (of varying strictures and with varying success) that encourage abortions after the
third pregnancy, and impose quasi-compulsory sterilization, delay
marriages, and — of late — try to push for one-child families. Images of noble restraint have often been projected, as with idealization
of the Red Guards' disciplined behaviour during the 'Cultural
Revolution' (during 1966–9), or reminders of the traditional Chinese
delay in sexual liaisons in comparison to the 'wanton sexuality of
the capitalist world' (Orleans, 1972 p. 39). Such ideals are hardly
conducive to undermining the (largely Confucian) predisposition to
have sons (*see* the newspaper *Xinhua*, 29 June 1979).

Nationalism presents a new threat of population because there is
always the danger that, in its bellicose expressions, appeals will be
made for extra manpower to militate against neighbouring states.
This looming problem applies particularly to black Africa (understandably so, considering the ravages of AIDS) and to Eastern
Europe. In Russia, for example, with its recent shifts to the Right,
we can expect to hear some pre-revolutionary views again — such
as those of Nikolai Mordvinov, who linked increased personal
wealth of elites to the increased number of workers (Valenty, Ed.,
1980 p. 85).

Nationalist policies can also take on an introverted face, so that
'each country shuts itself up within its own self-interest', discount-
ing global problems (Al-Hakim, 1978 p. 65). The wisdom of
religious leaders, and of intellectuals with globalist and/or spiritual
insights, is also likely to be neglected in favour of short-term political goals. Recent Australian positions on population matters, for
instance, have it that Australia herself has no population problem,
that she stands outside the sphere of the relevant global predicament,

justifiably maintains her current immigration rate, and sees no strain being imposed by her population policy on the natural environment — all views that well-known Australian academic globalists with religious outlooks would question (*see e.g.* Birch, 1993 esp. pp. 142–3).

What of other surrogates and quasi-religions? The attitudes of other than those 'implicit creeds' that are of a more intellectual orientation (*e.g.* humanism and Darwinism) are generally for population control but usually balance their approval with a concern for human rights and a distaste for socio-political manipulations. Some idiosyncratic espousers of eco-Darwinism can also be fatalist-looking, not seeing why humans could be destined to a power so special that they should be enabled to survive before other species (some Authors even defending the parental choice of infanticide, *cf.* Khuse & Singer, 1985 pp. 98–139). Other 'popular attitudinal' profiles can militate against an easier solution. So-called 'cargoism', to illustrate, is a mode of expectation about happiness, and along with the requisite new goods, images of a 'decent sized' family become part of a 'mind-set' — especially as television drama typically sets goodly amounts of 'the Cargo' and the typical family of five idylically living together.

INVOLVEMENT FOR SOLUTIONS

Reflecting on this survey, it is quite unrealistic to expect from religions any real panacea. They are divided between and within themselves, and the world faiths are constantly at work against pressures that are contrary to fulfilment of their highest ideals — including those of self-restraint (which are crucial for decreasing reproduction). On the other hand, *there remains immense, untapped potential in religions to meet crises.* Concerted religious instruction about social matters have been shown by demographic research to modify behaviour (Catholic and Protestant campaigns significantly reducing the size of adherent families in the Low Countries, for example, in the decade 1958–68, *see* Tabah, 1971 pp. 45–9). Religions have an extraordinary capacity to move people's consciences and reorient their lives to work for a better future (and this possibility will be vital for combating irresponsibilities towards both numbers and the environment — unless, of course, religious groups choose only to serve their own rather than common interests, or are 'captured' by narrowly nationalistic concerns).

There is certainly more room for collaboration between religious leaderships. Thus 1993 was the Year of Indigenous Peoples; yet it was also (as most seemed to forget) the Year of Interreligious Understanding and Cooperation, with a Parliament of Religions being held once more. These are incentives for further dialogue, mutual respect, and *cooperation*. The various environmental agencies around the world that are affiliated with religions have a golden opportunity to share their data, policies, and basic concerns, with regard to overpopulation. The achievements of the WWF show this to be possible, although the UN needs to consider how its secretariats and other bodies could coordinate religious efforts towards global readjustments far more effectively than they currently do. Medical professionals, who are also representatives of different credal and ideological persuasions, could and should be brought into a closer dialogical and working relationship (through WHO, UNEP, IUCN, and international environmental networks) than currently prevails. The world is not only overpopulating but is practically 'bleeding to death', *inter alia* through misunderstanding between traditions, and both problems can be addressed by concerted efforts at inter-faith (not merely international) cooperation.*

REFERENCES

AKBAR, K.F. (1994). Family Planning and Islam: A Review. *Hamdard Islamicus*, 17(3), pp. 85–94.
AL-HAKIM, T. (1978). Unforseen Challenges. Pp. 63–6 in *Suicide or Survival? The Challenge of the Year 2000* (A.M. M'Bow et al.). UNESCO, Paris, France: 192 pp.
BAH, S.M. (1992). Mortality and morbidity in Muslim populations: a framework for the reduction of their factors. *Hamdad Islamicus*, 15(2), pp. 75–86.
BIEZANEK, A. (1964). *All Things New*. Pan Books, London, England, UK: 172 pp.
BIRCH, C. (1993). *Regaining Compassion: For Humanity and Nature*. University of New South Wales Press, Sydney, NSW, Australia: 251 pp.
BOUVIER, L.F. & RAO, S.L. (1975). *Socioreligious Factors in Fertility Decline*. Ballinger, Cambridge, Massachusetts, USA: 204 pp.
CAMPANELLA, M. (1993). Global Society, Global Problems and New Forms of Decision-Making. Pp. 52–63 in *Transition to a Global Society* (Eds S.B. IRAJAYMAN & E. LASZLO). One World Publications, Oxford, England, UK: 176 pp.
CHAPPLE, C.K. (1993). *Nonviolence to Animals, Earth, and Self in Asian Traditions*. (SUNY Series in Religious Studies.) State University of New York Press, New York, NY, USA: 146 pp.
CLUB OF ROME (1975). *The Limits to Growth*. New American Library, New York, NY, USA: 207 pp.
CRIBBS, A. (1992). Panel presentation [along with Australian Aboriginal thinker A. PATTEL-GRAY], Seminar on *'Racism, Population and Sustainability'*. Church of Christ General Synod, including Network for Environmental and Economic Responsibility, St. Louis, Missouri, USA: 22 pp.
DE LESTAPIS, S. — see LESTAPIS, S. DE.

* Thanks for special details are due to Shah Brahman, Ruth Lewin-Broit, Lucy Davey, Cong-Tam Dao, Raul Fernandez-Caliense, Bahram Soroush-Kabouli, Adnan Kasaminie, Wei-Ping Liu, Graeme Lyall, Mehravara Marzbani, Anne Pattel-Gray, Paulus Ryanto, and Saman Shritprajna.

DRAPER, E. (1972). *Birth Control in the Modern World: The Role of the Individual in Population Control*. Penguin, Harmondsworth, England, UK: 446 pp.

EHRLICH, P.R. & EHRLICH, A.M. (1990). *The Population Explosion*. Simon & Schuster, New York, NY, USA: 320 pp.

EKD — *see* Evangelical Church in Germany.

ENGELS, F. [1878–83] 1954. *Dialectics of Nature* (Transl. C. Dutt). Progress Publishers, Moscow, Russia: 403 pp.

EPPSTEINER, F. (Ed.) (1988). *The Path of Compassion: Writings of Socially Engaged Buddhism*. Parrallax Press, Berkeley, California, USA: xx + 220 pp.

EVANGELICAL CHURCH IN GERMANY (cited as EKD) (1994). *'How Many People can the Earth Systain?' Ethical Reflections on the Growth of the World Population*. (EKD Texte 49.) EKD Advisory Commission for Development Affairs, Hannover, Germany: 66 pp.

FAGLEY, R.M. (1960). *The Population Explosion and Christian Responsibility*. Oxford University Press, New York, NY, USA: xix + 260 pp.

FARRANDS, J.L. (1993). *Don't Panic, Panic!* Text Publishing, Melbourne, Victoria, Australia: 198 pp.

FISCHER, J. (Ed.) (1965). *Die Problematik der 'Geburtenregelung' in der Diskussion der Kirchen: Quellenmaterial zur Meinungsbildung*. Evangelische Kirche in Deutschland, Stuttgart, Germany: 128 pp. [A volume partly inspired by Fagley's researches].

HILGERS, T.W. & HORAN, D.J. (Eds) (1972). *Abortion and Social Justice*. Sheed & Ward, New York, USA: xxv + 328 pp.

HINNELLS, J. (1984). Current research in the study of living Zoroastrianism. Pp. 48–54 in *World Zoroastrianism: Souvenir Issue*. (The 1st World Conference on Zoroastrian Religion, Culture, and History, held in London 29 June to 1 July 1984.) World Zoroastrian Organization, London, England, UK: 111 pp.

IKADO, F. (1968). Trends and problems of new religions: religion in urban society. Pp. 101–17 in *The Sociology of Japanese Religion* (Eds K. MORIOKA & W.H. NEWELL). (International Studies in Sociology and Social Anthropology.) Leiden, The Netherlands: 145 pp.

INTERNATIONAL SOCIETY FOR EDUCATIONAL INFORMATION Inc. (cited as ISEI) (1993). *The Japan of Today*. Japan Echo Inc., Tokyo, Japan: 157 pp.

ISKANDER, N. (Ed.) (1974). *Simposium Kebijaksanaan Kependukukan di Indonesia, Pandaan-Jawa Timur, 17–20 Desember 1973*. Lembaga Demografi, Jakarta, Indonesia: 220 pp.

JACKSON, M. & TEAGUE, T. (1978). *Mental Birth Control*. Lawton-Teague Publications, Oakland, California, USA: 63 pp.

KITAGAWA, J.M. (1960). *Religions of the East*. Westminster Press, Philadelphia, Pennsylvania, USA: 351 pp.

KUHSE, H. & SINGER, P. (1985). *Should the Baby Live? The Problem of Handicapped Infants*. Oxford University Press, Oxford, England, UK: vvv + 228 pp.

KÜNG, H. (1972). *Infallible? An Enquiry*. Collins/Fontana, London, England, UK: 224 pp.

LESTAPIS, S. DE (1959). *La Limitation des Naissances*. SPES, Paris, France: 315 pp.

LORIMER, E. (1954). *Culture and Human Fertility*. UNESCO, Paris, France: 513 pp.

MCDOWALL, D. (1993). Dilemmas of the Jewish state. Pp. 643–62 in *The Modern Middle East: a Reader* (Eds A. HOURANI *et al.*). I.B. Tauris, London, England, UK: 697 pp.

MONTEFIORE, H. (1970). *Can Man Survive? The Question Mark and Other Essays*. Collins, London, England, UK: 224 pp.

NOONAN, J.T. (1986). *Contraception: A History of its Treatment by Catholic Theologians and Canonists*. Harvard University Press, Cambridge, Massachusetts, USA, 581 pp.

ORLEANS, L.A. (1972). *Every Fifth Child: the Population of China*. (The China Library.) Eyre Methuen, London, England, UK: 191 pp.

OSBORN, L. (1993). *Guardians of Creation: Nature and Theology in the Christian Life.* Apollos, Leicester, England, UK: 172 pp.

PALMER, S.J. (1986). Community and commitment in the Rajneesh Foundation. *Update: a Quarterly Journal on New Religious Movements,* **10**(4), pp. 3–15.

POLUNIN, N. (1980). *Growth Without Ecodisasters? Proceedings of the Second International Conference on Environmental Future (2nd ICEF), held in Reykjavik, Iceland, 5–11 June 1977.* Macmillan, London & Basingstoke, England, UK: xxvi + 675 pp., illustr.

SARASWATI, M.D. (1993). *Gokarunanidhi: In Defence of the Cow with All Compassion* (transl. I. RAM). Dayan & Ashram, Ajmer, India: 48 pp.

SARPONG, P.A. (1977). Aspects of an African world-view. Pp. 57–64 in *Religion, Mortality and Population Dynamics* (Proceedings of the Seminar on Moral and Religious Issues in Population Dynamics and Development, University of Ghana, Accra, Ghana, 31 March–4 April 1974) (Ed. J.S. POBEE). Unviersity of Ghana, Accra, Ghana: 270 pp.

SRINIVAS, M.N. & RAMASWAMY, E.A. (1977). *Culture and Human Fertility in India.* Oxford University Press, Delhi, India: 32 pp.

SULLOWAY, A.W. (1959). *Birth Control and Catholic Doctrine.* Beacon Press, Boston, Massachusetts, USA: xxiii + 257 pp.

TABAH, L. (1971). *Rapport sur les relations entre fécondité et la condition sociale et économique de la famille en Europe: leurs répercussions sur la politique sociale.* (2e Conférence Démographique Européenne, 31 Aug.–7 Sept. 1971.) Conseil de l'Europe, Strasbourg, France: 145 pp.

TROMPF, G.W. (Ed.) (1990). *Cargo Cults and Millenarian Movements.* (Religion and Society 29.) Mouton De Gruyter, Berlin & New York: xvii + 466 pp.

TROMPF, G.W. (Ed.) (1993). *Islands and Enclaves: Nationalisms and Separatist Pressures in Island and Littoral Contexts.* Sterling, New Delhi, India: xxxv + 379 pp.

TROMPF, G.W. (1994). *Payback: The Logic of Retribution in Melanesian Religions.* Cambridge University Press, Cambridge, England, UK: xx + 545 pp.

VALENTY, D.I. (Ed.) (1980). *An Outline Theory of Population* (Ed. D.I. VALENTY). Progress Publishers, Moscow, Russia: 308 pp.

VAYDA, A. (1976). *War in Ecological Perspective.* Plenum, New York, NY, USA: xiv + 129 pp.

VEITCH, J. (Ed.) (1995). *Can the World Survive? The Religions of the World Respond to the Environmental Crisis.* Quo Vadis, Wellington, New Zealand. [Not available for checking.]

VERZOSA, M.P. (1989). Population control for whom? *Breakthrough* (Global Educ. Associates), **10**(2–3), pp. 65–6.

WORLD HEALTH ORGANIZATION (cited as WHO) (1984). Policies and programmes affecting mortality and health. Pp. 1–11 in *Mortality and Health Policy.* United Nations, Geneva, Switzerland: 200 pp.

WORLD WILDLIFE FUND (cited as WWF) (1990). Population in crisis: statements on behalf of various religions presented in *The New Road: the Bulletin of the WWF Network on Conservation and Religion* (the Population Issue), Nr 16, Oct.–Dec. 1990, pp. 1–3.

YUVAL-DAVIS, N. (1989). National reproduction and the 'Demographic Race' in Israel. Pp. 92–109 in *Woman – Nation – State* (Ed. N. YUVAL-DAVIS). Macmillan, London, England, UK: Houndmills, Basingstoke, England, UK: 185 pp.

12

Energy for a Sustainable World Population

by

HE PROFESSOR JOSÉ GOLDEMBERG,
Instituto de Electrotécnica e Energia, Cidade Universitária,
Av. Prof. Almeida Prado 925,
Universidade de São Paulo,
05508-900 São Paulo, Brazil;
Former *Minister of Education* and (interim)
Minister of Environment,
Brazil, during UNCED 1992; *Visiting Professor 1993–94,*
Woodrow Wilson School of Public & International Affairs,
Princeton University, Princeton, NJ 08544, USA;
formerly *Rector, University of São Paulo*

INTRODUCTION

A transition to a safe and sustainable world will require actions to overcome:

a) the *physical constraint* of guaranteeing enough material resources — such as food, water, and energy — for an increasingly demanding human population;

b) the *population constraint* which increases availability of existing resources *per caput*;

c) the *economic constraint* of adjusting to increased costs of material resources; and

d) the *political constraint* of guaranteeing a minimum of equity in access to adequate resources of food, water, energy, cultivable and building land, etc.

The 'physical' and 'economic' constraints to sustainability can probably be solved through technological advances. The 'political' constraints have plagued Mankind for thousands of years, and have led frequently to wars of conquest and slavery. Now, however, the emergence of democratic regimes in most parts of the world brings hope of attenuating or even dispelling such problems. Yet what underlines all of the above constraints is the continuing population growth which in many cases is faster than the production of an adequacy of goods and services, so frustrating efforts to improve the living conditions of the population at large and any attempted global equity.

235

One of the questions we will try to answer here is the role of energy as an instrument for socio-economic development and particularly for a reduction in the growth-rate of human population. In this context, energy is of little interest in itself. However, it is an essential ingredient of development and economic growth, the objective of the energy system being to provide *energy services*. These are the desired and useful products, processes, or services, that result from the use of energy — for instance in assuring illumination, a comfortable indoor climate, refrigerated storage, transportation, appropriate temperatures for cooking, etc.

Low energy consumption is not the cause of poverty. However, it is an indicator of many elements thereof, such as poor education, bad health-care, the 'status' of women in society, and the hardship imposed on women and children by poverty. What the empirical evidence indicates is that an increase in energy services is strongly correlated with a decrease of the total fertility-rate, in the sense of the number of children born per woman. Such perceptions seem to have been incorporated in the population debate, as witnessed by what happened in the Cairo Conference in 1994 (Senanayake, 1995).

The empowerment of women and their increased access to education, as well as an increased awareness of the need of governments' policies to avoid a population explosion, are indeed vital elements in the search for a sustainable world.

In our discussion we will also highlight the fact that the new problems resulting from global environmental degradation might actually help towards finding solutions to the political problems, as this is turning out to be in the self-interest of the industrialized countries as well as in the interests of the elites in developing countries as affecting all, rich and poor alike — *see*, for example, Houghton *et al.* (1990).

THE PHYSICAL CONSTRAINT

From a physical point of view, sustainable development means three things, namely:
1. The increased use of *renewable* natural resources such as agri-cultural products and timber. As far as energy is concerned, this means the increased use of renewable energy resources such as hydro-, solar-, and wind-power, instead of fossil fuels (Larson *et al.*, 1986; Reddy & Goldemberg, 1990; Goldemberg, 1997);
2. A stable population; and
3. Conservation and rational use of non-renewable materials and resources and, whenever possible, recycling.

In addition to the above, there are of course other dimensions to the problem of sustainable development. Consumption patterns and/or different types of societal organizations could help enormously in achieving the goal of sustainability. We will not discuss these possibilities here, but instead concentrate on what can be done to improve the sustainability of the 'present system'.

As the 'present system' we understand civilization at the point it has reached today, with great affluence in the OECD (Organization for Economic Cooperation and Development) countries and great poverty in the so-called LDCs (Less-developed Countries*). Eastern European Countries and the ex-Soviet-Union — often called 'non-OECD Europe' — are going through rapid changes and will probably follow the path of the OECD countries in the fairly near future.

In the OECD countries there are less than one thousand million people with an income *per caput* higher than approximately seven thousand US dollars per year and an energy consumption of some six metric tons (tonnes) of oil equivalent *per caput* annually (Goldemberg *et al.*, 1988). Energy consumption trends are quite representative of consumption of most other natural resources (Larson *et al.*, 1986). Population has stabilized in OECD countries as well as in 'non-OECD' European countries (World Population Prospects, 1993).

In the developing countries, income *per caput* is in general around one-tenth of what it is in industrialized countries and the energy consumption is less than one ton of oil equivalent *per caput* (Goldemberg *et al.*, 1988; Reddy & Goldemberg, 1990). This is clearly insufficient to provide a reasonable standard of living for the majority of the population in these countries, at least with present-day technologies.

The 'present system' is therefore not a sustainable one. Energy consumption is bound to increase, and in addition, the less-developed countries' population is growing at a rate of approximately 2% per year (World Population Prospects, 1993). Altogether, total energy consumption (and the use of most other natural resources) is growing at approximately 4% per year (British Petroleum Company, 1993) and is predicted to double in 17 years.

Fig. 1 shows what we are to expect in the first few decades of the next century as far as energy is concerned. The LDCs' overall consumption is predicted to surpass OECD's consumption already around the year 2010 at 4% per year growth or around AD 2020 at 3% per year growth (World Development Report, 1992).

If the sources of supply in both OECD and LDCs remain the same as they are today, in the next 20 years great problems will be

---

* For a listing of 'developing countries' *see* World Development Report (1992).

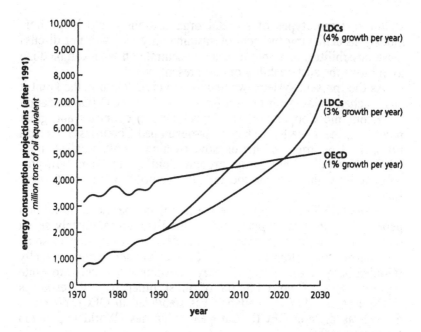

FIG. 1. *Projections of primary energy consumption, AD 1991–2030.*

faced by the industrialized countries (World Energy Council, 1993). These problems are particularly likely to be:

1. Ever-increasing competition for sources of energy supply in a period when oil and natural-gas reserves will inevitably be dwindling.

2. Severe regional and global environmental problems especially in LDCs due to the extensive use of coal (mainly in China and India, which are the most populous countries by far). Coal currently represents 39.4% of the total energy supply in LDCs and only 21.8% in the OECD countries (World Energy Council, 1993). In all likelihood, developed countries will be held at ransom by what happens outside their borders, at least in the environmental arena.

What can be done to face such threats?

In the first place we must introduce and practice methods of using energy more efficiently than currently — especially in the developing countries. To some extent one has learned how to do it in the industrialized countries and their experience should be transferred to LDCs — not merely as a matter of philanthropy, but also as a matter of self-interest as emphasized above. The same end-use services can be performed with less primary energy, *i.e.* burning less coal or oil and therefore reducing environmental damage (Goldemberg *et al.*, 1988).

However, the way to do it in developing countries is not through 'energy conservation', because energy consumption *per caput* is small in such countries and there is not much to conserve, but instead through 'technological leapfrogging'. The growth of energy consumption in developing countries is essential, due to the need of building the infrastructure in industry, transportation, urban development, and the like. As LDCs are growing, the extensive use of better technologies — including energy conservation — has to be incorporated earlier in the process of development and not as a 'retrofit' as happened in the industrialized countries. This is why they are important theatres for innovation and 'leapfrogging' — especially in the energy-intensive basic-materials industry, whereas demand has reached saturation in the industrialized countries (Reddy & Goldemberg, 1990).

Amazing indeed is the speed of adoption and diffusion of innovative and state-of-the-art technologies as developing countries modernize: when villages in India are electrified there will be no reason for not using fluorescent lamps instead of inefficient light-bulbs; colour TV is everywhere, with black-and-white TV already a thing of the past even in remote areas of Amazonia! In some countries, such as Vietnam, a telephone service is being introduced with wireless telephones ('cellular' telephones) which do away with the need of an extensive and cumbersome wiring system. Other examples of 'leapfrogging' solutions in the energy sector are (*cf.* Goldemberg, 1994):

1. The alcohol programme in Brazil, which produces 250,000 barrels of ethanol per day, so replacing one-half of the gasoline that would otherwise be used in the country, and generating 700,000 rural jobs. Being a renewable fuel, the use of alcohol has reduced the carbon dioxide (CO_2) remaining in the atmosphere from the emissions of Brazil by 18%* (*see also* Macedo, 1992).

2. Photovoltaics, which could play an important role in the tropical areas — where most of developing countries are located — not only in decentralized but also in centralized units. While such 'PV' technology is perhaps the most inherently attractive of the renewable technologies, it is also about the farthest from being commercial in bulk in the power market. By it, costs would be brought down quickly *via* mass purchases that could be facilitated by various national and international organizations in conjunction with increase in research and development (R&D) in the industrialized countries.

---

* Presumably owing to reabsorption of this equivalent in the process of photosynthesis constituting the basis of the alcohol used as fuel. — Ed.

3. Technologies for large-scale application of wind-power based on the exploitation of good wind resources that are remote from areas having high electricity demands. Key technologies, besides wind turbines, are compressed-air energy storage and long-distance DC (direct current) transmission lines. This combination could supply 'baseload wind electricity' at the end of the transmission lines and have attractive economic aspects.

4. Conversion of the military jet-engine industry to aeroderivative gas-turbine industry for stationary power-generation.

5. Development of fuels suitable for use in fuel-cell vehicles, such as hydrogen or methanol from natural gas (present technology) or from biomass *via* thermochemical gasification (with gasifiers that need to be demonstrated commercially). Initially it is very likely that natural gas would be used and converted near the point of use (*e.g.* at centralized bus depots in urban centres).

Other, more futuristic, schemes such as a 'hydrogen economy' have been proposed but it seems premature to consider them at the same level as the ones mentioned above. Both OECD countries and LDCs still rely heavily on the use of fossil fuels that will inevitably be exhausted even if used with far more wisdom than they are today. The solution to this problem is to use them more efficiently than at present, while gradually switching to the use of renewables such as wind- or solar-power.

TABLE I

Sources of Energy in Two Scenarios
(in Gtons* of oil equivalent).

YEAR AD 2020 WEC scenario C[†] (ecologically driven)			YEAR AD 2025[§]	
	Solid	2.1	Coal	2.00
	Liquid	2.7	Oil	1.72
	Gas	2.4	Gas	2.10
	Nuclear	0.4	Nuclear	0.33
			Geothermal	0.04
Renewables 3.4 (30%)	Large Hydro	0.9	Hydro	0.68
	'New' Renewable[‡]	1.4	Intermittent Renewables	0.84
	'Traditional'**	1.1	New biomass	3.30
			Solar H$_2$	0.20
	Renewables 5.02 (45%)			
TOTAL		11.3		11.21

* 1 Gton = 10⁹ tons; † *cf.* World Energy Council (1993); §*cf.* Johansson *et al.* (1993); ‡ 'New' renewables = solar, geothermal, modern biomass, ocean, and small hydro; ** 'Traditional' biomass = fuelwood, crop residues, and dung.

A number of studies have been conducted to find out the extent of what is possible in this area, and the best means of attaining such goals. Table I shows the results of two of these studies (Johansson *et al.*, 1993; World Energy Council, 1993).

The World Energy Council (WEC) scenario (1993) predicts that, by the year 2020, approximately 30% of all energy in use in the world could come from 'traditional' and 'new' *renewable* resources. The study by Johansson *et al.* (1993) predicts that, by the year 2025, 45% of the energy used could come from renewable sources such as hydro-power, intermittent renewables — wind-power, photovoltaics, etc. — and biomass through modernized methods or hydrogen produced by the electrolysis of water using solar energy.

These advances, if made, would not take us to a fully sustainable energy future but would point the way to further progress to be achieved in the next century.

There are, therefore, reasons for some optimism regarding the achievement of the goal of sustainable development as far as energy and materials are concerned, except for the problem of recycling non-renewable materials. This task remains as one of the great challenges for the next few decades. Progress in recycling of urban wastes is taking place in several countries with the active parti-cipation of the public, which is very encouraging. The use of organic wastes for electricity and heat generation is taking place in several European countries and in the United States. Recycling automobiles and other domestic appliances will be, however, one of the main problems in the future, because sometimes they become harder to recycle and relatively more costly as the technology employed improves. This is clearly the case with high-performance motors in which ceramic materials are used.

THE POPULATION CONSTRAINT

One attractive feature of an increase in energy-use in developing countries — with higher efficiency than at present — is that it leads to a decrease in the total fertility rate (TFR)* and therefore lessens the basic problems of population growth (Goldemberg, 1994). The connection between energy and fertility is illustrated in Fig. 2. This figure shows how the TFR decreases dramatically with energy consumption in almost all countries, with some exceptions mainly in the Middle East and Africa.

One should mention here that what counts for people is not so much the amount of energy spent in performing given tasks as the

---

* TFR is the total number of children born to each woman. If TFR = 2, the population ceases to grow.

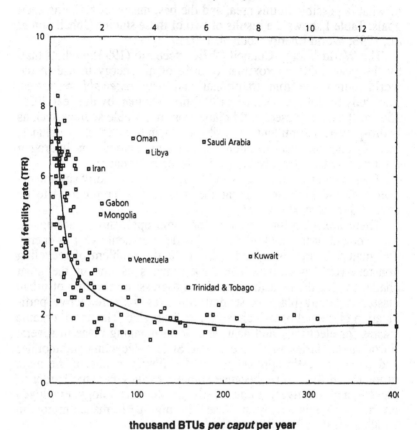

FIG. 2. *Total fertility rate as a function of* per caput *energy consumption (all countries).*
From Goldemberg (1994).

services obtained from that energy. If the conversion efficiency of the energy input is low (primary energy), more energy is needed to perform the same task.

Fig. 3 shows in greater detail than Fig. 2 the relationship between total fertility rate and yearly energy consumption *per caput* for some countries in which this consumption is below 3 kw.

Although there is considerable scattering of points in each of these figures, some features can be established with reasonable confidence:

a) Most Muslim countries have fertility rates that are very high when compared with others having approximately the same energy consumption *per caput;*

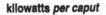

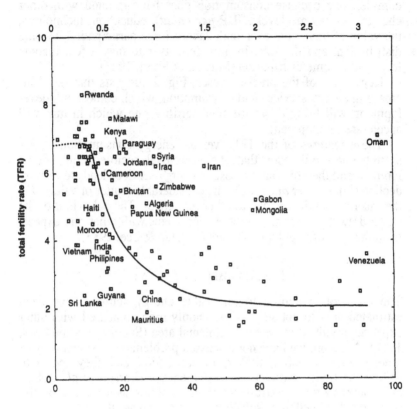

FIG. 3. *Total fertility rate as a function of* per caput *energy consumption (LDCs).*
From Goldemberg (1994).

b) Countries such as Sri Lanka (and Bangladesh) have very low fertility rates compared with others having approximately the same energy consumption *per caput;*

c) Excluding the latter category of countries, one has a reasonable fit to the points: for countries with consumption *per caput* higher than 1 kw, TFR changes rather smoothly and is approximately constant around 2; and

d) For the developing countries (*cf.* Fig. 3) it is possible to envisage a fit through most points, which indicates a rise in TFR with income followed by a rapid decrease.

The relationship between energy consumption and fertility rate is a complex one because consumption goes hand-in-hand with other changes — income level (GDP *per caput*), education, technology, the value of time, culture, female labour-force participation, income distribution, and the like. In turn, fertility rate may affect income level and income distribution (Mueller & Short, 1983).

Regardless of the precise causes, Fig. 2 suggests that rapid increase in energy services and consumption, which comes with development, will bring down the total fertility rate which in turn will accelerate development.

These features of the TFR 'versus' energy consumption curves seem to confirm the view that, at a macrolevel, socio-economic development and therefore an increase in energy services leads to fertility decline (Entwisle *et al.*, 1982). The considerable scatter of values, *i.e.* the macrovariability, is due to other influences. The behaviour of the curve drawn through the points is consistent with what one expects from theories of population growth (Bulatao & Lee, 1983).

THE ECONOMIC CONSTRAINT

The costs of switching to renewable energy sources have been estimated and do not seem to be unduly high compared with other expenses involved in the environmental area (Stavins & Whitehead, 1992). Although we have not discussed problems of sustainability of other resources such as timber, minerals, food, etc., they appear to follow the trends outlined above for energy (Larson *et al.*, 1986), and there seem to exist technological solutions to most of the problems that need to be solved to attain sustainability.

Just to give one example, the cost of preventing or adapting to the 'greenhouse effect' has been estimated as roughly 2% of the total world gross domestic product (GDP) (Stavins & Whitehead, 1992). Considering the fact that, today, 100 thousand million dollars are spent in the United States alone on environmental protection, it does not seem to be too far-fetched an idea to imagine that the governments and populations of developed countries will accept the extra burdens of moving to a sustainable future in this respect.

In the particular case of energy, the low cost of oil, at present, makes it difficult to persuade people of the importance of the adoption of alternative solutions based on renewables. However, a doubling in the cost of oil in the next 25 years would certainly change this picture drastically, as was demonstrated by the oil crisis of the decade AD 1970–80, caused by the abrupt rise in oil prices. An example is given by the Alcohol Programme in Brazil. The best available estimates place the cost of ethanol currently at approximately US $30 per barrel-equivalent of gasoline (Goldemberg *et al.*,

1993). With the decline of the cost of petroleum, the Alcohol Programme in Brazil faced difficult problems, and the solution found by the Government to preserve it was to index the consumer price of alcohol to the price of gasoline — which locally is, traditionally, approximately double its price to the consumer in the United States. The proceedings of this 'tax' on gasoline are used to reduce the cost of other petroleum derivatives, namely Liquified Petroleum Gas (LPG) and diesel oil. In the case of alcohol, a similar 'tax' is imposed on gasoline to alleviate the higher production costs of alcohol.

The justifications for such a policy are the beneficial environmental and social consequences of the Programme.

Other general studies on the costs of achieving the goal of 'sustainable development' have been conducted by the Secretariat of the United Nations Conference on Environment and Development (UNCED June 1992) and took the form of *Agenda 21*, a rather comprehensive document which evaluates how much it would cost to move into a sustainable future (Robinson, 1993). The answer is that some 125 thousand million dollars per year would represent the needed financial flow from the industrialized countries.

This is approximately double the present amount spent as ODA (Official Development Aid) by the developing countries. To convince governments of the industrialized countries to do it seems rather problematical, at least currently, because of the general economic crisis in the OECD countries. Their pressing problems have received and are receiving higher priorities than long-term strategies which are inherent in any attempt to achieve a sustainable future. There is also some misunderstanding as to what are the real environmental problems and who is interested in eliminating them. Meanwhile it is useful to recognize that the main categories of pollution are the following (Goldemberg, 1993):

Local pollution is that which is intimately linked to underdevelopment, such as the lack of clean water and air, and failure of services of removal and disposal of solid wastes and liquid effluents. What in effect characterizes poverty in the slums of great cities is the fact that the population lives mainly among the rubble and residues which it produces, due to the lack of the resources needed to build sewers and other engineering works. In rural areas, the poor degrade the environment by overexploiting it through deforestation often leading to ultimate desertification.

Regional pollution is caused mainly by automobiles and heavy industry, which are inherent in the more prosperous societies. Huge cities of more than 10 million souls — such as Los Angeles, Mexico City, and São Paulo — have long been 'suffering' from pollution caused by the emissions (mainly particulates, sulphur, and nitrogen oxides) resulting from the burning of fossil fuels, and from 'smog'. Sometimes the amount of pollution produced is large enough to cause regional and even transborder problems, such as the 'acid rain' originating ultimately in the United States but responsible for the destruction of life in Canadian lakes. The same has happened to lakes in Scandinavia, due to industrial activities mainly on the other side of the Baltic Sea.

The transport 'choking' in such cities as Los Angeles, São Paulo, Santiago, Mexico City, and others, forces the rich to work closer to the rest of the population than they would otherwise do, and so they all suffer from the same environmental problems.

Global pollution constitutes the third category, and its most alarming consequences so far are the partial destruction of the stratospheric ozone 'shield' by chlorofluorocarbons (CFCs) and the warming of the globe produced by gases such as CO_2 which are a result of anthropogenic activities (Houghton *et al.*, 1990). These problems result from changes in the composition of the atmosphere and have little to do with national borders. The causes of such global problems are gases originating anywhere in the world, and are such that, for example, the well-being of people living in New York might ultimately be affected by what takes place in India or China (and *vice versa*).

As far as local pollution is concerned, local governments can expect little help. The rich countries have been unusually callous about problems of local pollution, except for loans offered for amelioration through such international concerns as The World Bank. However, as funds available from such concerns are limited, the authorities bestowing them and especially recipients have to make often agonizing choices between fighting pollution or using the loans for developmental purposes.*

In the case of regional pollution, usually few countries are involved and they ought to be able to sort out the problems among themselves. This was done for example between Germany and Poland, when the East German Government decided to help Poland by paying for filters in their smoke-stacks which were sending pollutants across the borders into Germany.

The developing countries contribute, at this time, only modestly to the emissions of chlorofluorocarbons and 'greenhouse' gases; but

---

* Ironically this reminds us of an item of personal micro-history that has plagued us through much of our life and become known in our immediate circle as 'Polunin's syndrome'. It emanated from an occasion in our youth when a member of a committee (we think it was of the Royal Society of London deciding on a limited award although it may have been elsewhere towards one of our arctic expeditions) told us that he was sorry they had not given it to us but when they had almost decided to do so somebody had made the point that they had better give it to another as 'Polunin will do his stuff anyway'. — Ed.

as their aggregate populations keep on growing, as well as their economies, their 'contributions' are bound to increase as is pointed out above.

THE POLITICAL CONSTRAINT

The technological and economic problems involved in achieving a sustainable future thus do not seem to be insurmountable. The problem resides in the political will to do so.

Within the natural boundaries of today's so-called developing countries — where three-quarters of the world population lives — one finds usually an 'elite' who not only hold the political power but also maintain a very comfortable and sometimes extravagant standard of living. The elite rich are generally surrounded by a majority of the people who do not have access to the amenities that, nowadays, are considered normal or even esssential for a reasonable standard of living (Goldemberg *et al.* 1988; Reddy & Goldemberg, 1990). The image of 'islands of affluence surrounded by a sea of poverty' characterizes reasonably well such cities as Nairobi, São Paulo, Jakarta, and many others.

In developing countries, *development* means satisfying the basic human needs of all the population — including access to jobs, food, health services, education, housing, running water, sewage disposal, etc. The lack of access to such services by the majority of citizens, and the fact that only a minority of the people have access to them, is the fertile ground of political unrest and revolutions. More than that, the hopelessness and despair which leads people to emigrate to the industrialized countries has its origins in such all-too-widespread situations.

Achieving sustainability requires bridging the gap between rich and poor nations and, within individual countries, bridging the gap between the elites and the masses. This is what politics is really about, and the bridging of such gaps will only take place if the rich countries (especially the elites within those countries) perceive that it is in their long-term self-interest to do so and act accordingly.*

The case for self-interest in the matter of global environmental degradation, *i.e.* 'greenhouse' warming and related effects, seems to be compelling, and this is probably what drove so many countries to agree on the need of a Climate Convention in the UNCED 1992 Conference in Rio.

In other areas, such as guaranteed access to oil and other exhaustible and sometimes fast-being-used-up resources, little agreement exists and the richer countries seem determined to protect their

---

* A sobering example is the fact that melting down of the Antarctic Ice-cap alone to the present sea-level would raise the world sea-level sufficiently to inundate most capital cities. — Ed.

privileged access to them. Equity considerations will presumably play only a small role in this area as they have always done in the past. Only a further aggravation of existing problems, as has happened in a number of instances, will generate the political will that is needed to embark on lasting solutions.

ACKNOWLEDGEMENT

I would like to close with a note of warm tribute to the initiator and Editor of this timely anthology whose exemplary treatment has been such that I have been happy to approve each and every one of his corrections and suggestions which have made at least my chapter more readable in its final stage for publication.

REFERENCES

BRITISH PETROLEUM COMPANY (1993). *B.P. Statistical Review of World Energy* (June 1993). Britannic House, 1 Finsbury Circus, London EC2M 7BA, England, UK: 37 pp., illustr. and tables.
BULATAO, R.A. & LEE, R.D. (1983). An overview of fertility determinants in developing countries. Chapter 25, vol. II of *Determinants of Fertility in Developing Countries* (Eds R.A. BULATAO et al.). Academic Press, New York, NY, USA: 2 vols, illustr. and tables.
ENTWISLE, B., HERMALIN, A.I. & MASON, W.M. (1982). *Socioeconomic Determinants of Fertility Behavior in Developing Nations: Theory and Initial Results*. National Academy Press, Washington, DC, USA: 125 pp., illustr. and tables.
GOLDEMBERG, J. (1993). The geopolitics of environmental degradation. *Environmental Conservation*, 20(3), pp. 193–4.
GOLDEMBERG, J. (1994). *Energy for Development*. (Center for Energy and Environmental Studies, Report Nr 283.) Princeton University, Princeton, NJ 08544, USA: 199 pp., illustr. and tables.
GOLDEMBERG, JOSÉ (1997). *Energy, Environment, and Development*. Earthscan Publications, 120 Pentonville Road, London N1 9JN, England, UK: 174 pp., illustr. and tables.
GOLDEMBERG, J., JOHANSSON, T.B., REDDY, A.K.N. & WILLIAMS, R.H. (1988). *Energy for a Sustainable World*. Wiley Eastern, India: 517 pp., illustr. and tables.
GOLDEMBERG, J., MONACO, J. & MACEDO, I.C. (1993). The Brazilian fuel-alcohol program. Pp. 841–63 in *Renewable Energy — Sources for Fuels and Electricity*. Island Press, Washington, DC, USA: 884 pp., illustr. and tables
HOUGHTON, J.T., JENKINS, G.J. & EPHRAUMS, J.J. (Eds) (1990). *Climate Change — The IPCC Scientific Assessment*. Cambridge University Press, Cambridge, England, UK: 884 pp., illustr. and tables.
JOHANSSON, T.B., KELLY, H., REDDY, A.K.N. & WILLIAMS, R.H. (Eds) (1993). *Renewable Energy — Sources for Fuels and Electricity*. Island Press, Washington, DC, USA: 884 pp., illustr. and tables.
LARSON, E.D., ROSS, M.H. & WILLIAMS, R.H. (1986). Beyond the era of materials. *Sci. Am.*, 254(6), pp. 34–41, illustr. and tables.
MACEDO, I.C. (1992). The sugar cane industry: its contribution to reducing CO_2 emissions in Brazil. *Biomass and Energy*, Vol. 3, Nr 2, pp. 77–80.
MUELLER, E. & SHORT, K. (1983). Effects of income and wealth on the demand for children. Chapter 18, Vol. I of *Determinants of Fertility in Developing Countries* (Eds R.A. BULATAO et al.) Academic Press, New York, NY, USA: 2 vols, illustr. and tables.

REDDY, A.K.N. & GOLDEMBERG, J. (1990). Energy for the developing world. *Sci. Am.*, **258**(9), pp. 111–8, illustr. and tables.

ROBINSON, N.A. (Ed.) (1993). *Agenda 21: Earth's Action Plan, Annotated.* (IUCN Environmental Policy and Law Paper Nr 27.) Oceana Publications, New York, NY, London & Rome: xciv + 683 pp.

SENANAYAKE, P. (1995). Reflections on the UN's International Conference on Population and Development. *Environmental Conservation*, **22**(1), pp. 4–5.

STAVINS, R.N. & WHITEHEAD, B.W. (1992). Pollution charges in environmental protection: a policy-link between energy and environment. *An. Rev. Energy and Environment*, **17**, pp. 187–210.

WORLD DEVELOPMENT REPORT (1992). *Development and Environment.* The World Bank — Oxford University Press, Oxford, England, UK: 308 pp., illustr. and tables.

WORLD ENERGY COUNCIL (1993). *Energy for Tomorrow's World.* Kegan Paul, London, England, UK: [not available for checking].

WORLD POPULATION PROSPECTS (1993). *The 1992 Revision.* United Nations, New York, NY, USA: 667 pp., tables.

13

Facing Nature's Limits

by

LESTER R. BROWN, MS (Maryland), MPA (Harvard)
President, *Worldwatch Institute,*
1776 Massachusetts Avenue, NW,
Washington,
DC 20036 - 1904,
USA.

INTRODUCTION

In September 1994, the 179 national delegations assembled in Cairo at the United Nations' International Conference on Population and Development, reached agreement on a plan designed to stabilize the world's human population. This World Population Plan of Action may well prove to be the boldest initiative ever undertaken by the United Nations, dwarfing some of its earlier achievements, such as even the eradication of smallpox. On the twenty-eight anniversary of his 1969 first landing of a man on the Moon, we can paraphrase the American astronaut Neil Armstrong and conclude that this initiative of the UN Conference in Cairo was a giant step for Humankind.*

In the preparatory meetings leading up to the Cairo Conference, delegates had rejected the notion that population growth would continue on the high trajectory, reaching 11.9 thousand millions by AD 2050. Instead, they opted for an extraordinarily ambitious plan to stabilize population between the medium projection of 9.8 thousand millions in AD 2050 and the low projection according to which population would peak at 7.9 thousand millions by AD 2050. The latter, vastly preferable strategy reflects a sense of urgency — a feeling that, unless population growth can be slowed very soon and effectively, it will push human demands beyond the carrying capacity of the land and aquatic resources in many countries, leading to environmental degradation, economic decline, and ultimately social disintegration.

---

* See *'Cairo Conference Adopts 20-Year Program of Action Linking Population, Development and Women's Empowerment'*, press release, International Conference on Population and Development Secretariat, New York, NY, USA, 30 September 1994; UN General Assembly, *Programme of Action of the United Nations International Conference on Population and Development* (draft), New York, 19 September 1994; also Senanayake (1995).

In the second half of the nineteen-nineties, evidence that the
world is on an economic path which is environmentally unsustainable or worse can be seen in shrinking fish-catches, falling water-
tables, declining bird-populations, record heat-waves, and dwindling
grain-stocks — to name just a few of the many situations giving rise
to concern or downright alarm.

The world fish-catch, which had climbed more than fourfold
during the recent 40 years (1950 to 1989), is no longer rising,
apparently because oceanic fisheries cannot sustain a greater catch.
The failure to coordinate population policy with earlier carrying-
capacity assessments of fisheries means that the world now faces a
declining seafood supply per person and generally rising seafood
prices for the future.*

Concern over scarcity of fresh water is rising in many areas. A
prolonged drought in northern China, for example, and the associated
water shortages, have raised questions about the suitability of Beijing
as the national capital, and renewed discussion of a 1,400-kilometres
(860-miles-long) canal that would bring water from the south to the
water-deficient north.

Although collapsing fisheries and water scarcity attract attention
because of their immediate economic effects, the decline of bird
populations may be a more revealing indicator of the Earth's
declining health. Recently-compiled data by BirdLife International,
of Cambridge, England, show that populations are declining on every
continent. Thus of the world's 9,600 species of birds, only 3,000 are
holding their own in numbers, the other 6,600 being in decline. Of
these, the populations of some 1,000 species have dropped to the
point where they are threatened with extinction.

After two decades of steadily-rising global average temperature,
including the highest on record in 1990, the June 1991 eruption of
Mount Pinatubo in the Philippines gave the world a brief respite from
global warming. The eruption ejected vast amounts of sulphate
aerosols into the upper atmosphere, which soon spread around the
globe. When once in the upper atmosphere, the aerosols reflect a
minute amount of incoming sunlight back into space — enough,
however, to exert a cooling effect. Yet by early 1994, almost all the
aerosols had settled out, clearing the way for a resumption of the
warming trend.[†]

Thereafter, evidence of new temperature highs was not long in
coming. A pre-monsoon heat-wave in central India lasted several

---

* Fish-catches from UN Food and Agriculture Organization (FAO), *Fishery
Statistics: Catches and Landings* (Rome, Italy: various years), and from FAO, Rome,
Italy, private communications, 20 December 1993.

† Effects of Mount Pinatubo on atmospheric temperatures, from 'Climate Group
Rejects Criticism of Warnings': *Nature* (London), 22 September 1994.

weeks, with temperatures up to 46 degrees Celsius (115 degrees Fahrenheit) taking a heavy toll on humans and livestock in the region. For the western United States, hundreds of new records were set, creating hot, dry conditions that led to a near-record number of forest fires.*

Japan had its hottest summer on record in 1994. Intense heat led to excessive evaporation and water shortages so severe that many utilities and manufacturing firms in Tokyo and surrounding areas were forced to import water by tanker from as far away as Alaska. Over a thousand miles (1,600 km), to the south-west of Tokyo, Shangai — with little air-conditioning — suffered during July through 14 days above 35 degrees Celsius (95 degrees Fahrenheit) and another 16 days between 33 and 34 degrees Celsius. Moreover in parts of Northern Europe — including Germany, Poland, and the Baltic states — mid-summer temperatures soared well above 32 degrees Celsius, exposing both residents and ecosystems to unaccustomed levels of heat.†

On the food front, developments were particularly disturbing. Even though in 1994 the United States returned to production of all the grainland that had been idled under commodity supply management programmes, global food security declined further as the world's projected carry-over grain-stocks from the 1994 harvest dropped to the lowest level in 20 years. A combination of increasing shortages of water, declining fertilizer-use, and cropland losses — particularly in Asia — led to another harvest shortfall and the drawdown in stocks.§

Thus in various ways, Nature's limits are beginning to impose themselves on the human agenda, initially at the local level, but in time also at the global level. Some of these impositions, such as the yield of oceanic fisheries or extending water-scarcity, are near-term. Others, such as the limited capacity of the atmosphere to absorb excessive emissions of carbon without disrupting climate, will manifest themselves over the longer term.

* 'India Sweltering Through Deadly Heat, Drought', *Los Angeles Times*, 31 May 1994; Michael McCarthy, 'USA: New Fires Worry Beleaguered Crews in West', *Reuter Newswire*, 5 August 1994; 'Firefighters Battling 33 Major Blazes in the West', *Reuter Newswire*, 10 August 1994.

† 'Japanese Utility to Import Water As Heat Wave Saps Refineries', *Journal of Commerce*, 17 August 1994; Gordon Cramb, 'Drought-hit Japan is to Import Water', *Financial Times*, 18 August 1994; Peter Blumberg, 'Japan Water Deal Fulfills Alaskan Officials' Dream', *Journal of Commerce*, 30 August 1994; 'Heat Wave', *China Daily*, 3 August 1994; 'Summer's a Scorcher Across the Atlantic, Melting Old Records', *Wall Street Journal*, 5 August 1994.

§ U.S. Department of Agriculture (USDA), Economic Research Service (ERS), *'Production, Supply, and Demand View'* (electronic database), Washington, DC, USA, November 1996.

THREE IMMINENT LIMITS

One of the key questions that emerged as the world prepared for the 1994 Cairo Conference (*see* my opening paragraph) was, *How Many People can the Earth Support?* Closely-related was: What Exactly will Limit the Growth in Human Numbers? Will it be the scarcity of water, life-threatening levels of pollution, food scarcity, or some other limiting condition? After considering all the possible constraints, it appears to us that the supply of food will most likely determine the Earth's population carrying-capacity. Three of the Earth's natural limits are already slowing the growth in world food production: the sustainable yield of oceanic fisheries, the amount of fresh water produced by the hydrological cycle, and the amount of fertilizer that existing crop varieties, and their habitat support, can effectively use.

More than 20 years have passed since a marine biologist at the UN Food and Agriculture Organization (FAO) estimated that oceanic fisheries could not sustain an annual yield of more than 100 million tons. In 1989, the world fish-catch, including that from inland waters and fish-farming, reached exactly that total — an amount approximately equal to the world's production of beef and poultry combined. During the following four years this total catching or otherwise harvesting of fishes fluctuated between 97 million and 99 million tons, reducing the fish-catch per person 8% in these four years. Recent FAO reports indicate that all the 17 major oceanic fisheries are now being fished at or beyond their capacity. With the total catch unlikely to rise much above 100 million tons, the decline in the seafood etc. supply per person of the last few years will continue indefinitely— or at least until the World Population Plan of Action succeeds in stabilizing population.*

A combination of pollution and overharvesting is 'killing' many inland seas and coastal estuaries. The Aral Sea, for instance, once yielded 44,000 tons of fish per year; but latterly wholesale diversion to irrigation of its supplying rivers' water has greatly shrunk the Aral Sea, raising its salt content and making the salt in effect a pollutant. All the 24 species of fish that were once fished there commercially are believed to be extinct so far as the Aral Sea is concerned. In the Caspian Sea, the famous sturgeon (*Acipenser* sp.) harvest has been reduced, through pollution and over-fishing, to perhaps 1% of the level of 50 years ago (Postel, 1992).

---

*The estimate of 100 million tons is from the FAO-sponsored publication, J.A. Gulland (Ed.) *The Fish Resources of the Ocean*, Fishing News Ltd., Surrey, England, UK, 1971) — *see also* Gulland (1993), who indicates a 'current annual harvest from the sea of 80 million tonnes; World Resources Institute, *World Resources 1994–95*, Oxford University Press, New York, NY, USA 1994); FAO, *Fishery Statistics: Catches and Landings* (Rome, Italy: various years), with updates from FAO as private communications.

The Black Sea, which is the dumping receptacle for the Danube, Dniester, and Dnieper, Rivers, is the repository for chemical and organic pollutants for 'half of Europe'. Of the nearly 30 species that once supported commercial fisheries there, only 5 remain. During the last decade the total annual catch has dropped from nearly 700,000 tons to 100,000 tons — a result of pollution, overharvesting, and the accidental introduction of destructive alien species of fish.*

The US Chesapeake Bay, once one of the world's most productive estuaries, is deteriorating rapidly from a lethal combination of pollution, overharvesting, and — in oysters — disease. Formerly a major source of this delicacy, the Bay's annual harvest has dropped from nearly 100,000 tons of edible oysters (roughly 1 million tons in the shell) around the turn of the century to less than 1,000 tons in 1993[†] (see Fig. 1).

Thousand Tons

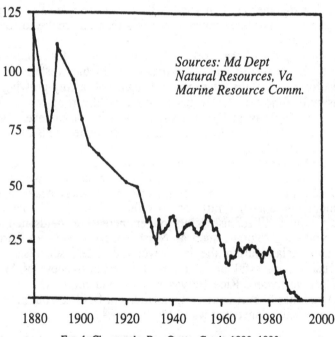

Sources: Md Dept
Natural Resources, Va
Marine Resource Comm.

FIG. 1. Chesapeake Bay Oyster Catch, 1880–1993.

* John Pomfret, 'Black Sea, Strangled by Pollution, Is Near Ecological Death', *Washington Post*, 20 July 1994.

† Conversion figure from FAO, Fishery Information, Data and Statistics Service, 'Conversion Factors — Landed Weight to Live Weight', Rome, Italy, March 1992; Figure 1 from John Jacobs, Maryland Department of Natural Resources, 'Eastern Oyster, Fishery Statistics of the United States' (unpublished printout), April 1994, and from Virginia Marine Resource Committee, 'Oyster Ground Production' (unpublished printout). Newport News, Virginia, USA, April 1994.

With land-based food-stocks, limits on production are being imposed by the amount of fresh water supplied by the hydrological cycle. Today, two-thirds of all the water extracted from rivers and underground aquifers is used for irrigation. In parts of the world where all available water is now being used, such as the southwestern United States or large areas of Northern China, satisfying future growth in residential and industrial demand will have to come at the expense of agriculture (Postel, 1992), further exacerbating the food-scarcity problem.

Although there are innumerable opportunities for increasing irrigation efficiency, only limited potential exists to expand fresh-water supplies for irrigation. For example, roughly one-fifth of US irrigated land is watered by 'drawing down' (i.e. depleting) under-ground aquifers. A recent study of India found that water-tables are now falling in several states, including much of the Punjab (India's bread-basket), Haryana, Uttar Pradesh, Gujurat, and Tamil Nadu — states that together contain some 250 million people. The drop in underground water-level ranges from less than one metre to several metres a year.[*]

In many parts of the world, the diversion of water to non-farm uses is also reducing water for irrigation. In the western United States, for instance, the future water demands of rapidly-growing Las Vegas will almost certainly be satisfied by diverting water from irrigation. Similarly in China, most cities suffer from severe water shortages, and many of them will meet their future needs by taking water away from irrigation.[†]

The physiological limit on the amount of fertilizer that current crop varieties can use poses an even broader threat to world food expansion. In countries where fertilizer use is already heavy, applying more nutrients has little or no effect on yield. This helps to explain why fertilizer use is no longer increasing yields in major food-producing regions, such as North America, Western Europe, and East Asia. During the last several decades, scientists were remarkably successful in increasing the responsiveness of Wheat (*Triticum aestivum*), Rice (*Oryza sativa*), and 'corn' (Maize, *Zea mays*) varieties to ever-heavier applications of fertilizer, but in recent years such efforts have met with little success.[§]

---

[*] US situation from Dickason (1988), and Clifford Dickason, USDA, ERS, Washington, DC, USA, private communication, 19 October 1989; Malik & Faeth (1993); Population Reference Bureau (PRB), Washington, DC, USA, private communication, 5 October and 17 October 1994.

[†] Timothy Egan, 'Las Vegas Stakes Claim in 90s Water War', *New York Times*, 10 April 1994; Patrick E. Tyler, 'China Lacks Water to Meet Its Mighty Thirst', *New York Times*, 7 November 1993.

[§] Duane Chapman & Randy Barker, *Resource Depletion, Agricultural Research, and Development* (Cornell University, Ithaca, NY, 1987); Fertilizer Industry Association (IFA), *Fertilizer Consumption Report* (Paris, France, 1992).

World-wide, fertilizer use increased tenfold between 1950 and 1989, when it peaked and then began to decline. During the following four years it fell some 15%, with the decline concentrated in the former Soviet Union following the withdrawal of subsidies. In the United States, fertilizer use peaked in the early nineteen-eighties and has declined by roughly one-tenth since then. In China, the other leading food-producer, the peak seems to be occurring roughly a decade later. Some countries, such as Argentina and Vietnam, can still substantially expand their use of fertilizer, but the major food-producing countries are close to the limit with existing grain varieties (IFA, 1992; *Fertilizer Yearbooks*, FAO, Rome, Italy, various years).

For nearly four decades, steadily rising fertilizer use was the engine driving the record growth in world food output. The generation of farmers on the land in 1950 was the first in history to double the production of food. By 1984, they had outstripped population growth enough to raise *per caput* grain output an unprecedented 40%. But when the use of fertilizer began to slow down in the late nineteen-eighties, so did the growth in food output (IFA, 1992; *cf. Fertilizer Yearbook*, FAO, Rome, Italy).

The era of substituting fertilizer for land-use to increase grain production came to a halt in 1990 (*see* Fig. 2). If future food output gains cannot come from using large additional amounts of fertilizer, where will they come from? As the twenty-first century approaches, the graph of fertilizer use and grainland area per person may epitomize the human dilemma more clearly than any other depiction could do. The world has quietly, with very little fanfare, entered a new era — one fraught with uncertainty over how to feed the projected massive growth in world population*.

Unless plant breeders can develop strains of wheat, rice, and corn, that are much more responsive to fertilizer than those currently used[§], the world may not be able to restore the rapid growth in grain output that is needed to keep up with human population growth. Either science will have to come up with a new method of rapidly expanding food production or otherwise satisfying human nutritional needs (*e.g.*[§]), or population levels and dietary patterns will be forced to adjust to much tighter food supplies. With the prospect of no growth in ocean-based food supplies and of much slower growth henceforth in land-based food supplies project major, the world is facing a future far different from the recent past.

---

* Fig. 2 from IFA (1992), and from US Bureau of the Census projections, published in Francis Urban & Ray Nightingale, *World Population by Country and Region, 1950–1990, with projections to 2050* (Washington, DC, USA: USDA, ERS, 1993).

[§] Or others can devise some totally different system for nourishing humans than anything at present known — as seems conceivable in the face of such recent wonders as cloning mammals. — Ed.

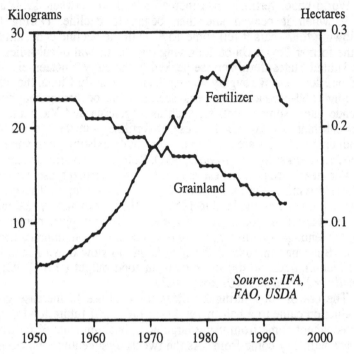

FIG. 2. World Fertilizer Use and Grainland Area Per Person, 1950–94

THE ECONOMIC EFFECTS

The depletion of natural capital — of forests, rangelands, topsoil,
underground aquifers, and fish stocks, not to mention minerals and
fossil fuels — and the pollution of air and water, have reached the
point in many countries where the economic effects are becoming
highly visible, including a loss of output, of jobs, and of exports.
Some countries have lost entire industries.

As the global demand for seafood overruns the sustainable yield
of fisheries, or as pollution destroys their productivity, for instance,
fisheries collapse — raising seafood prices, eliminating jobs, and
shrinking the economy. The economic wreckage left in the wake of
these collapses can be seen around the world: fishing villages that
once lined the Aral Sea are now mere 'ghost towns', with their
modern craft stranded far up on dry land (*see Environmental Conservation*, **20**(2), p. 192, 1993). In Newfoundland, the collapse of the
Cod (*Gadus morhua*) and Haddock (*Melanogrammus aeglefinus*)
fishery has left 33,000 fishers and fish-processing workers unemployed, crippling the province's economy. And in New England,

families who for generations have made their living from the sea are selling their trawlers and searching for other employment (Postel, 1992).*

Even as fisheries are being destroyed, the world demand for seafood is rising. Seafood was once a cheap source of protein — something that people ate when they could not afford meat. In 1960, a kilogram of seafood cost only half as much as a kilogram of beef. In recent years, however, that margin has narrowed and finally disappeared, as seafood prices have risen above that of beef. During the last decade the world price of seafood, in real terms, has risen nearly 4% a year (*see* Fig. 3).†

Dollars Per Ton

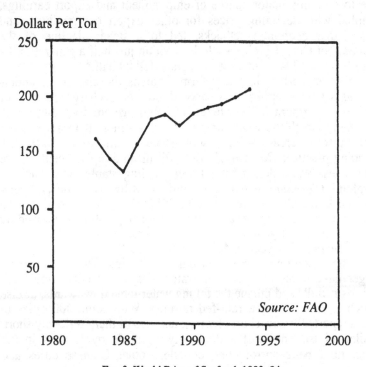

FIG. 3. World Price of Seafood, 1983–94

* See also Anne Swardson, 'Canada Closes Section of Atlantic to Fishing', *Washington Post*, 10 April 1994; Christopher B. Daly, 'Fishermen Beached As Harvest Dries Up', *Washington Post*, 3 March 1994.

† Beef prices from US Department of Labor, Bureau of Labor Statistics, 'Consumer Price Index' (unpublished printout), Washington, DC, USA, 21 April 1994; Fig. 3 from FAO, *Fishery Statistics: Trade and Commerce* (Rome, various years), with updates from Adele Crispoldi, Fishery Statistician, Fishery Information, Data and Statistics Service, Fisheries Department, FAO, Rome, Italy (unpublished printout), 12 September 1994. World fish price calculated by dividing total global catch by total global value, as opposed to recording prices from any one location.

In some economies, over-cutting forests has done even more economic damage than over-fishing. The clear-cutting of tropical hardwood forests by lumber companies has almost completely destroyed this valuable resource in some developing countries, devastating their economies. Côte d'Ivoire, for example, enjoyed a phenomenal economic expansion in the nineteen-sixties and -seventies as its rich tropical hardwood forests yielded export earnings of $300 millions a year. It thus became a 'development model' for the rest of Africa; but, as in many other countries that did not practise sustainable forestry, clear-cutting decimated its forests, and so exports dropped to $30 millions a year in the early nineteen-nineties. The loss of this major source of employment and export earnings, coupled with declining prices for other export commodities, and some other economic setbacks, led to a steady decline of the economy of Côte d'Ivoire, such that, within just half a generation — from 1980 to 1994 — income per person fell by half.*

As noted earlier, in many farming areas the claims on underground water supplies now exceed aquifer recharge rates. For farmers in northern India, where wheat and rice are double-cropped, the rate at which the water-table is falling — more than a metre per year in some areas — may soon force a shift to less-intensive cropping practices. Most likely this will mean a replacement of rice with a less water-demanding, lower-yielding staple crop, such as Sorghum (*Sorghum vulgare*) or Indian Millet (*Oryzopsis hymenoides*). Although this may arrest the fall in the water-table, it is not a welcome development in a country whose population is expanding by 17 millions per year and is projected to reach a thousand millions within the next few years (Malik & Faeth, 1993).†

In the agricultural regions surrounding Beijing, China, farmers no longer have access to reservoir water. They must now either drill their own wells and pursue the falling water-table downwards, or else switch to less intensive rain-fed farming. With some 300 cities in China reportedly now short of water, and 100 of them seriously short, similar adjustments will undoubtedly be made by farmers in the agricultural belts surrounding countless other Chinese cities and towns.§

In the southwestern United States, the need to supply booming cities with water, together with the depletion of aquifers, is leading to the diminution of irrigated agriculture in many areas. In arid Arizona,

* FAO, '*Time Series for State of Food and Agriculture 1993*' (electronic database), Rome, Italy, 1993; Steve Coll, '*Sub-Saharan Africa's Incredibly Shrinking Economies*', *Washington Post*, 6 August 1994.

†*See also* Population Reference Bureau, Washington, DC, USA: *1994 World Population Data Sheet*.

§ Patrick E. Tyler, 'China Lacks Water to Meet Its Mighty Thirst,' *New York Times*, 7 November 1993.

the diversion of irrigation water to the rapidly-growing 'sunbelt' cities of Phoenix and Tucson means that large areas of hitherto productive farmlands have returned to desert. In the Texas 'panhandle', where the southern reach of the Ogallala Aquifer has been largely depleted, farmers have reverted to dryland farming. Although agriculture continues in this region, the drop in intensity, and hence of output, reduces employment in both the agricultural input and service industries as well as in the agricultural processing industries. As a result, some rural communities are being partially depopulated (Postel, 1992).

In situations where years of overpumping is depleting aquifers, reductions in irrigation perforce lie ahead. If the rate of ground-water pumping in an area is double the rate of recharge, for example, the aquifer will eventually be depleted. As it nears depletion, the withdrawal rate will necessarily be lowered more and more, because it cannot for long exceed the recharge rate, and so the irrigated area will be reduced accordingly.

In contrast to Wheat and Maize crops, which are largely rain-fed, the production of Rice depends heavily on irrigation. This makes yield-trends easier to analyse, simply because the effect of weather fluctuations is much more than with irrigation.

An even more telling indicator of the loss of momentum in expanding grain output is the drawdown in world stocks since 1987. Then, world carry-over stocks of grain from the 1985 harvest totalled 465 million tons — an all-time high and equivalent to 104 days of consumption. During the following ten years, grain stocks were reduced to 254 million tons — a drop of 211 million tons or some 21 million tons a year (*see* Fig. 4). This annual drawdown exceeds the yearly growth in the world grain-harvest during this period, which averaged roughly 10 million tons. Stated otherwise, a substantial part of the growth in world grain consumption since 1987 comes from consuming stocks — a trend that cannot continue much longer, because current stocks represent only 51 days of consumption, or less than in-the-pipeline supplies (Table 1).

In response to the position of decline in grain stocks, in 1993 the United States released for production in 1994 all grainland that had been idled under its commodity supply management programmes. Even with this land returned to use, and even with fair-to-excellent growing conditions in the world's major food-producing regions, stocks continue to decline. If grain stocks cannot be rebuilt in years of good harvests, when will they be rebuilt?*

---

* Idled cropland data from assorted USDA press releases, May–July 1994 and various years, scarcely help to answer this question.

Days

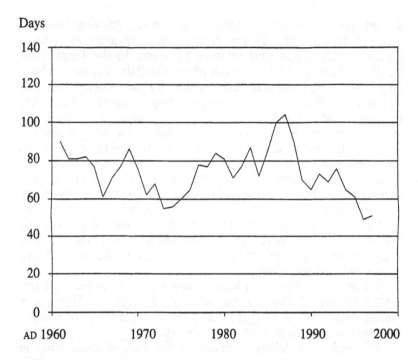

FIG. 4. World Grain Carry-over Stocks as Days of Consumption, 1961–96. Compiled by Worldwatch Institute.

In the absence of any dramatic new technological advance in agriculture comparable with the discovery of fertilizers or the hybridization of Maize, there is now a real possibility that grain production could continue to lag and that prices could begin to rise in the years ahead, in the manner of those of seafood as already indicated. The unfortunate reality is that, with carry-over stocks at such a low level, the world is now only one poor harvest away from chaos in world grain markets.

The collision between continuously expanding human demands and Nature's various limits, affects not only the world food supply but also overall economic growth. A 1993 study published by the World Bank notes that environmental damage takes many forms, including land degradation, pollution damage, the loss of biological diversity, deforestation, and soil depletion or erosion. Using a dozen or so examples, the two Authors — both economists — show that the annual costs to different countries of various forms of environmental damage can range from less than half of 1% (USA and Mali) to as much as 17.4% (Nigeria) of their gross national product (see Table II). If the data were available to calculate all the economic costs of

TABLE I

World Grain Carry-over Stocks, Quantity and as Days of Consumption, 1961–97.

Year	Quantity million tons	Consumption (days)
1961	203	90
1962	182	81
1963	190	81
1964	193	82
1965	194	77
1966	159	61
1967	190	71
1968	213	77
1969	244	86
1970	228	76
1971	193	62
1972	217	68
1973	180	55
1974	193	56
1975	200	60
1976	220	65
1977	280	78
1978	278	77
1979	328	84
1980	315	81
1981	288	71
1982	309	77
1983	357	87
1984	304	72
1985	366	85
1986	434	100
1987	465	104
1988	409	90
1989	316	70
1990	301	65
1991	342	73
1992	317	69
1993	351	76
1994	313	65
1995	296	61
1996	237	49
1997	254	51

Note: Data are for year to when new harvest begins.
Source: US Department of Agriculture, 'World Grain Situation and Outlook' (unpublished printout), Washington, DC, various years.
See Worldwatch publications *Vital Signs 1996* and *State of the World 1996* for further information.

environmental degradation in its many forms, they would undoubtedly show an enormous loss.

The exceeding of sustainable yield thresholds in sectors such as forestry and fishing, and of aquifers as we have already seen,

TABLE II

Estimates of Environmental Damage in Selected Countries to Dates Indicated.*

Country and Year	Form of Environmental Damage	Annual Costs as a Share of GNP
		%
Burkina Faso	Crop, livestock, and fuel-wood, losses from land degradation	8.8
Costa Rica (1989)	Deforestation	7.7
Ethiopia (1983)	Effects of deforestation on the supply of fuel-wood and crop output	6.0–9.0
Germany (1990)†	Pollution damage (air, water, soil pollution, loss of biodiversity)	1.7–4.2
Hungary (late 'eighties)	Pollution damage (mostly air pollution)	5.0
Indonesia (1984)	Soil erosion and deforestation	4.0
Madagascar (1988)	Land-burning and erosion	5.0–15.0
Malawi (1988)	Lost crop production from soil erosion Costs of deforestation	1.6–10.9 1.2–4.3
Mali (1988)	On-site soil erosion and losses	0.4
Netherlands (1986)	Some pollution damage	0.5–0.8
Nigeria (1989)	Soil degradation, deforestation, water pollution, other erosion	17.4
Poland (1987)	Pollution damage	4.4–7.7
United States§ (1981)	Air pollution control	0.8–2.1
(1985)	Water pollution control	0.4

*SOURCE: 'Environmental Damage Robs Countries' Income', *World Bank News*, 25 March 1993, based on David Pearce & Jeremy Warford; *Toward a World Without End* (World Bank, Washington, DC, USA, 1993).

† Federal Republic of Germany before unification.

§ Measures the benefits of environmental policy (avoided rather than actual damages).

combined with the slow-down in the growth in world grain production, directly affects the performance of the world economy and casts doubt on its durability and very viability. To begin with, these primary-producing sectors play a unique role in the global economy. If the growth in production of food from both land- and ocean-based sources falls far behind the growth in demand, the resulting rise in prices could destabilize some national economies.

Although projections by the International Monetary Fund (IMF) show economic growth accelerating in the years immediately ahead, these could be derailed by the instability associated with food scarcity. The nineteen-nineties could turn out to be the first decade since the Great Depression when income per person for the world as a whole actually declines in real-terms of purchasing value. Incomes fell in some 53 countries containing more than 800 million people during the nineteen-eighties, many of them in Africa. But that incomes might fall for the entire world during the 'nineties has not been anticipated in any long-range economic projection.*

UNSUSTAINABILITY FEEDS INSTABILITY

When once the demand for a particular product, such as seafood, exceeds the sustainable yield of its resource-base, any traditionally stable relationship between demand and supply becomes unstable. With thresholds for sustainable yields now being exceeded for so many resources, relationships that have been stable for centuries or millennia are becoming highly volatile late in the twentieth century.

Analysis of the relationship between the level of human demand and the sustainable yield of various systems is severely handicapped in many cases by a lack of data. For example, it is known that water-tables are falling in many countries because water-pumping now exceeds aquifer recharge. As noted earlier, some 21% of irrigated cropland in the United States depends on 'drawing down' underground aquifers. But for most of the world, data on sustainable aquifer yields are not available. Few communities or countries know when rising water-demand will exceed aquifer recharge; over-pumping is often discovered after the fact.†

We are also handicapped in analysing the effects of excessive demand on natural systems by the interaction of biological, economic, and political, systems. While academic specialists may understand the workings of individual systems and how they respond to stress, they seldom comprehend their interactions.

At some point, severe ecological stresses begin to manifest themselves economically on a scale that has political consequences. Rwanda is one tragic example of this. While the attention of the world in the summer of 1994 focused on the tribal conflict between the Tutsis and the Hutus, there was ample evidence that tensions had been building as the relationship between all Rwandans and the natural systems on which they depend deteriorated.

Between 1950 and 1994, Rwanda's population increased from 2.5 to 8.8 millions. The average number of children per woman in 1992, namely 8, was the highest in the world. Despite impressive gains in overall grain production, output per person declined by nearly half between 1960 and the early nineteen-nineties. Land scarcity intensified as increasingly small plots were subdivided from one gener-

---

* IMF, *Annual Report 1994* (Washington, DC, USA, 1994); World Bank, *World Development Report 1992* (Oxford University Press, New York, NY, USA, 1992).

† Dickason (1988), and Clifford Dickason, private communication, 19 October 1989.

ation to the next. As the population grew, the freshwater supply per person dropped to the point where Rwanda was officially classified by hydrologists as one of the world's 27 water-scarce countries.*

Beyond these numbers, however, was the quiet desperation that comes to an agrarian society when population growth overwhelms the carrying capacity of the land. Just as a lightning strike in forests in the American West is more likely to turn into an uncontrollable conflagration when the weather is unbearably hot and dry, so too are ethnic conflicts more likely to erupt when there are underlying tensions about food and the ability to earn a living.

Another essentially agrarian economy where the situation is in some ways even worse is that of Haiti. Once richly forested, it has lost all but 2% of its forests and much of its topsoil. In contrast to Rwanda, where the overall harvest has continued to rise, grain production in Haiti was one-third less in the early nineteen-nineties than it was in the mid-seventies, which means that grain production per person has plummeted. Political scientist Thomas Homer-Dixon observes that 'the irreversible loss of forests and soil in rural areas deepens an economic crisis that spawns social strife, internal migration, and an exodus of "boat people".' He concluded that even when President Aristide was returned to power, 'Haiti will forever bear the burden of its irreversibly ravaged environment, which may make it impossible to build a prosperous, just and peaceful society' (Homer-Dixon, 1994).

The ecological symptoms of unsustainability include shrinking forests, thinning soils, receding aquifers, collapsing fisheries, expanding deserts, and rising global temperatures. The economic symptoms include economic decline, falling incomes, rising unemployment, price instability, and a loss of investor confidence. The political and social symptoms include human hunger and malnutrition, and, in extreme cases, mass-starvation; environmental and economic refugees; social conflicts along ethnic, tribal, or religious, lines; and riots and insurgencies. As stresses build-up on political systems, governments weaken — losing their capacity to govern and to provide basic services, such as police protection. At this point, the nation-state disintegrates, replaced by a feudal social structure governed by local warlords — as in Somalia, now a nation-state only in name.

One of the difficulties in dealing with the complex relationship between humans and natural systems is that, when once rising

---

*Population from Bureau of the Census (Urban & Nightingale, 1993); fertility rate from Population Reference Bureau, Washington, DC, USA; agricultural statistics from USDA, Economic Research Service (ERS), *Production, Supply, and Demand View* (electronic database), Washington, DC, USA, July 1994; water situation from Postel (1992).

demand for seafood or firewood has crossed the sustainable yield threshold of fisheries or forests; future growth is often maintained only by consuming the resources-base itself. This combination of continuously rising demand and a shrinking resource-base can lead from stability to instability and thence to collapse almost overnight.

When sustainable yield thresholds are exceeded, the traditional responses proposed by economists no longer work. One common reaction to scarcity, for instance, is to invest more in production. Thus the key to alleviating seafood scarcity is to invest more in fishing trawlers. But in today's world this only exacerbates the scarcity, hastening the collapse of the fishery. Similarly, as food prices rise, there is a temptation to spend more on irrigation. But where water-tables are already falling, investing in more wells simply accelerates the depletion of the aquifer and the eventual decline in irrigation.

When the demand on a particular system reaches a limit, the resulting scarcity sometimes spills over to intensify pressure on other systems. As seafood became scarce, for example, many expected that fish-farming would take up the slack. But maintaining the historical growth in seafood supplies of 2 million tons per year over the last four decades by turning to aquaculture, where 2 kilograms or more of grain are needed to produce 1 kilogram of fish, requires at least 4 million tons of additional grain a year for fish raised in cages or ponds. Growth in the seafood harvest, which once relied primarily on spending more on diesel fuel to exploit ever-more distant fisheries, now depends on expenditures on grain as more fish are produced in marine feed. With grain supplies tightening, the feed may not be available to sustain rapid growth in aquacultural output.*

Some effects of crossing sustainable yield thresholds are indirect. If excessive demand for forest and livestock products leads to deforestation and rangeland degradation, the amount of rainfall runoff increases and with it erosion, while the amount of water retained and absorbed for aquifer recharge decreases. Thus, excessive demand for timber and livestock products can reduce aquifer yields.

As another example of an indirect effect, when carbon emissions exceed carbon fixation, as is happening with the massive burning of fossil fuels, the level of carbon dioxide in the atmosphere rises, altering the Earth's heat-balance. The principal effect is to trap heat, raising temperatures. This in turn affects all the ecosystems and ecocomplexes on which humans depend, from estuaries to range-lands.*

---

* Grain-to-fish conversion ratio from Robert Walters, 'Aquaculture Catches On'. *Mt Vernon Register News*, 31 July 1987.

Crossing sustainable-yield thresholds of natural systems can alter world markets. Ever since World War II, a challenge to agricultural policymakers, except for a brief period in the early-to-mid-nineteen-seventies, has been how to manage surpluses. Exporting countries typically insisted on using subsidies to bring farm prices above world market levels. This stimulated over-production, leading to the use of export subsidies to compete for inadequate import markets for grain. Now that production is no longer keeping up with growth in demand at current prices, policymakers may once again be faced with managing scarcity and dealing with the politics of scarcity as the historical decline of grain prices is reversed. This new trend is already evident in the seafood market.

Managing scarcity could test the capacity of national governments and international institutions. For example, overseeing fisheries was relatively easy when the catch was far below the sustainable yield. But when the catch exceeds that level, reestablishing a balance between the catch and the regenerative capacity of fisheries can be difficult. Similarly, countries that share water-basins find it relatively easy to manage water supplies when there is a surplus, but if water becomes scarce and there is no longer enough to go around, the problem of management increases inordinately.

The natural systems on which the economy depends — whether it be the hydrological cycle or forest or rangelands — are not merely sectors of the global economy, but its very foundation. If their productivity is diminished, then the prospect for the global economy will deteriorate. In an urbanized world where attention focuses on growth in telecommunications and computers and on the construction of the information superhighway, it is easy to forget that it is those natural systems which underpin the global economy.

One unfortunate and little-noticed consequence of these various trends of environmental and economic decline is that international assistance programmes are focusing more on aid and less on development. In effect, expenditures are shifting from crisis prevention to crisis management. Nowhere is this more evident than at the United Nations, where the budget for the UN High Commissioner for Refugees is nearly as high as that of the entire UN Development Programme. The same trend can be seen in Somalia, where social disintegration and conflict reached the point where military intervention was needed just to deliver the food supplies required to end famine. When physical and social deterioration descends to this

---

† cf. Intergovernmental Panel on Climate Change, *Climate Change: The IPCC Scientific Assessment* (Cambridge University Press, New York, NY, USA, 1990).

point, military intervention can easily cost 10 times as much as the food assistance which is being given.*

The bottom line of the growing instability between often overcrowded human societies and the natural systems on which they depend, is political instability. This in itself is beginning to make economic development and agricultural progress difficult, if not impossible, in many countries. In some countries, the exceeding of limits has international repercussions, an example of which is offered by China.

With its 1.2 thousand millions' population, China is facing a potential grain deficit so large that it could overwhelm the total exporting capacity of the United States and other exporting countries. The resulting fierce competition among importing countries will inevitably drive grain prices far above familiar levels.

The demand for grain in China is soaring as consumption of pork, poultry, eggs, beef, and beer, rises with income. With the economy expanding at 10% to 14% annually, China's record number of people are moving up the food-chain at a record rate. Never before have the incomes of so many people risen so rapidly, moving so many people up the food-chain at the same time. In addition, China is projected to add another 500 million people in the decades ahead, before its population stabilizes. Yet even as the record rate of industrialization raises living standards, extensive areas of cropland are being diverted to the construction of factories at a rate that is reducing grain production.

If the deficit develops as projected, food prices will rise throughout the world, affecting everyone. The inevitable outcome will be that when China turns to the world markets on an ongoing basis, its food scarcity will become the world's scarcity, and its shortages of cropland and water will become the world's shortages.

Contrary to popular opinion, it will probably not be in the devastation of such poverty-stricken countries as Somalia and Haiti, but through the booming economy of China, that we may feel most the inevitable collision between expanding human demands for food and the limits of some of the Earth's most basic natural systems. The shock-waves from this collision may be expected to reverberate throughout the world economy, with consequences that we can now only begin to foresee.

---

* UN High Commissioner for Refugees (UNHCR) budget from Heather Courtney, public information officer, UNHCR, Washington, DC, USA, private communication, 4 October 1994; UNDP budget from Adrianus de Raad, UNDP, New York, NY, USA, private communication, 19 October 1994.

REFERENCES *

BROWN, LESTER R. (1995). Nature's Limits. Pp. 1–20 in *State of the World 1995*, L.R. BROWN, C. FLAVIN, H.F. FRENCH, & L. STARKE (Editor), with 12 Contributing Researchers. W.W. Norton & Co, New York & London: xvi + 255 pp., figs & tables.

DICKASON, C. (1988). Improved estimates of groundwater mining acreage. *Journal of Soil and Water Conservation*, pp. 239–40.

FERTILIZER INDUSTRY ASSOCIATION (cited as IFA) (1992). *Fertilizer Consumption Report*, Paris, France: pp. 1–53 + Appendix.

GULLAND, J.A. (1993). Uses and abuses of the sea: fishing. Pp. 120–40 in N. POLUNIN & SIR JOHN BURNETT (Eds). *Surviving With The Biosphere: Proceedings of the Fourth International Conference on Environmental Future* (4th ICEF), held in Budapest, Hungary, during 22–27 April 1990. Edinburgh University Press, Edinburgh, Scotland, UK: xxii + 572, illustr.

HOMER-DIXON, T.F. (1994). Environmental scarcities and violent conflict: evidence from cases. *International Security*, Summer, pp. 5–40.

IFA — *see* FERTILIZER INDUSTRY ASSOCIATION.

MALIK, R.P.S. & FAETH, P. (1993). Rice–wheat production in northwest India. Pp. 17–32 in *Agricultural Policy and Sustainability: Case-studies from India, Chile, the Philippines, and The United States* (Ed. PAUL FAETH), World Resources Institute, Washington, DC, USA: 40 pp.

PEARCE, D. & WARFORD, J. (1993). *Toward a World Without End*. World Bank News, Washington DC, USA: XII (12), 40 pp.

POLUNIN, N. (Ed.) (1980). *Growth Without Ecodisasters? Proceedings of the Second International Conference on Environmental Future (2nd ICEF), held in Reykjavik, Iceland, 5–11 June 1977*. Macmillan, London & Basingstoke, England, UK: xxvi + 675 pp., figs & tables.

POSTEL, SANDRA (1992). *Last Oasis: Facing Water Scarcity*. W.W. Norton & Company, New York, NY, USA: 227 pp.

SENANAYAKE, P. (1995). Reflections on the UN's International Conference on Population and Development. *Environmental Conservation*, 22(1), pp. 4–5.

URBAN, F. & NIGHTINGALE, R. (1993). *World Population by Country and Region, 1950–1990, with projections to 2050*. US Bureau of the Census, Washington, DC, USA: p. 42.

---

* Many more, less complete or merely minor, 'newspaper' references, are given in the informative footnotes pervading this chapter. — Ed.

14

Hopes for the Future

by

SIR MARTIN W. HOLDGATE, PhD (Cantab.), F.I.Biol.
formerly: *Senior Biologist, British Antarctic Survey;*
Chief Scientist, UK Departments of Environment & Transport;
President of UNEP Governing Council;
Director-General of IUCN 1988–94:
35 Wingate Way, Trumpington, Cambridge CB2 2HD, England, UK,

&

Professor GAYL D. NESS, PhD (Michigan)
University of Michigan Population Fellow,
IUCN: The World Conservation Union,
Rue Mauverney 28,
1196 Gland, Switzerland

SIGNS OF HOPE

Despite its warnings of foreseeable ecodisasters, this book is, in itself, a sign of hope. It shows how, apart perhaps from conceivable major meteoric impact, even the worst could be averted by global concerted action in time. It avoids the hysteria and overstatement that so often characterize discussions of human population issues. It demonstrates that human population growth need not lead inevitably to environmental degradation. Moreover it emphasizes that human population questions have to be considered in the wider context of relationship between the human species and other species and components of the Biosphere, and also of the process of 'sustainable development', based on care for the environment. Finally, it demonstrates that Governments and people throughout the world are recognizing that rapid human population growth is a problem which must be addressed — and that more and more Governments are addressing it within the context of national social policy.

There are six particular signs of hope:

1) during the past quarter-century there has been a major advance in our understanding of how people interact with, depend on, and damage, the environment;

2) the leaders of the nations have come to recognize that sound develop-
    ment requires due care for the environment, and have committed them-
    selves to the process of 'sustainable development';
3) although environmental degradation remains widespread, and major
    global problems must be attacked urgently, corrective action is in hand
    in many regions;
4) the need for action to limit human population growth is now almost
    universally recognized;
5) practical ways of taking such action are now established, based on a
    blend of technical measures to enable people to limit their fertility and
    social measures to relieve poverty, enhance economic growth and
    personal opportunity (especially for women), and provide health-care
    and other supportive infrastructure*; and
6) most important of all, a demographic transition towards population
    stability is taking place in the developing world — at a faster rate than
    that previously experienced in the industrialized countries*.

UNDERSTANDING THE LIMITS AND CONSTRAINTS

'Humanity must live within the carrying capacity of the Earth.
There is no other rational option in the longer term. Unless we use
the resources of the Earth sustainably and prudently, we deny
people their future. We must adopt life-styles and development
paths that respect and work within nature's limits. We can do this
without rejecting the many benefits that modern technology has
brought, provided that technology itself works within those limits.'

This, the opening paragraph of the summary of *Caring for the
Earth: A Strategy for Sustainable Living* (IUCN/UNEP/WWF,
1991), defines the essential 'boundary conditions' for human development. It relates the numbers of people, their life-styles and
consumption patterns, and the benefits that technological skill can
bring, to the fundamental features of the Biosphere. 'Carrying
capacity' is here defined as the capacity of the Biosphere to support
the human species.

It is obvious that all species of biota are limited numerically by
the amount of space, energy, and other essentials, which they are
able to appropriate. Some, such as the Giant Panda (*Ailuropoda
melanoleuca*) or the Scottish Red Grouse (*Lagopus scoticus*), are
effectively dependent on a single food-genus or -species — in the
above examples, bamboos (*Bambusa* etc. spp.) and the Ling Heather
(*Calluna vulgaris*), respectively — and are inevitably liable to
population fluctuations if the abundance of these food-plants
changes markedly. But in virtually all cases, species' numbers are

---

* *See* especially Chapter 10 of this updated anthology. — Ed.

limited through a competitive process, arising because all animals are capable of producing more offspring than are needed to maintain population stability, the surplus being eliminated by competition for territory, food, mates, or other essentials. The human species is no different: indeed it has a rather high reproductive capacity for an animal of its size and longevity, and this has unquestionably in the past been beneficial by enabling it to expand rapidly into open habitats — for example following the glacial periods — and to recover quickly from famines or disease epidemics.

In today's world it is, however, evident that this high human reproductive capacity is ceasing to be beneficial. There are few empty habitats to occupy, or surpluses of food waiting to be consumed. It is no longer possible for surplus populations, or peoples displaced by conflict or famine, simply to migrate to unoccupied regions.

Compare, for example, the situation in Ireland in the mid-19th century and in Ethiopia or Somalia 130 years later. Around AD 1840, there were some eight million people living in Ireland, and they were heavily dependent on a single food resource — Potatoes (*Solanum tuberosum*), grown on very small family plots. In the mid-1840s the Potato Blight Fungus, *Phytophthora infestans*, invaded the island, spread rapidly, and devastated the Potato crop. In the succeeding five years, the human population crashed from eight millions to around five millions (Smith, 1962). About one million people died and two millions emigrated, mainly to North America. The emigrants were following a pattern that had become well-established among western European peoples. Between AD 1820 and 1930, over 50 million Europeans crossed the oceans, mainly to the Americas and Australasia (Crosby, 1986), dispossessing indigenous peoples and creating the societies we know today. But by the 1980s, this process had become increasingly impracticable. Social resistance to immigration is high, and still growing. The victims of famine in Somalia, Ethiopia, and many other countries, and the masses of poor people who see no prospect of a decent life in the lands of their birth, are effectively penned within their national frontiers or, if driven sufficiently by desperation to overspill them, are commonly confined within refugee camps on neighbouring territories which offer no prospect for long-term settlement. Pressures on land, competition for limited resources, and environmental degradation caused by attempts at cultivation and livestock husbandry on inappropriate soils, are boiling over into conflict in regions as far apart as Latin America and the Horn of Africa (Maguire & Brown, 1986; Hutchison, 1991).

Many species have evolved behavioural mechanisms that regulate breeding populations through non-lethal conflict for territory.

However, these mechanisms do lead to the displacement of unsuccessful competitors into 'marginal' habitats where survival is unlikely. The human species, on the other hand, can take social means that
curb population growth without involving the natural regulatory
mechanisms of starvation, aggression, and misery. Moreover the
human species can, and does, consciously enhance the 'carrying
capacity' of its environment, so that a given area supports more
people than formerly. The hope for the future lies in an interplay of
these two processes: the application of technological skills and
scientific understanding to increase carrying capacity, and the
strengthening of the technical and social processes which will
encourage and enable people to limit their fecundity.

INCREASING CARRYING CAPACITY

Carrying capacity can be defined as the number of people who can
be supported sustainably, for an indefinite period, by a particular
area of environment. As such the process is familiar to any agri-
culturalist, who knows very well the sustainable yield that is likely
to be derived by cropping a particular field with a particular application of fertilizer and pesticide, or the number of cattle, sheep, or
other grazing animals, that can be safely maintained on a particular
area of pasture. In historical times, the carrying capacity of an area
for people was closely related to its agricultural production, as
transport of food surpluses from area to area was difficult, and
generally speaking a population survived on what it could grow.

The process of development has been one of changing the
biological productivity of different regions of the planet for human
benefit. A very large number of sub-strategies have been involved in
this process, for example the following:

1) primary production of edible plants has been enhanced by selective
   breeding of the most productive or otherwise appropriate cultivars, by
   other biotechnological means, and by both the elimination of other plant
   species that compete with them ('weeds'), and the exclusion of
   herbivores through means as diverse as fencing, killing wild ones such
   as elephants and deer, and the application of insecticides;

2) yields have been further boosted by irrigation and by the application of
   compounds of nitrogen, phosphorus, and other essential nutrients,
   whether in the form of animal manures or synthetic fertilizers;

3) livestock breeding has produced strains of docile, quick-growing
   domestic animals, yielding meat, milk, and other products, at much
   higher rates than in the original 'founder' stocks that had been taken
   from the wild;

4) the predators that formerly consumed these domesticates have been
   widely eliminated, and veterinary science has protected the domesticates from a wide range of diseases; and

5) in a similar way, the management of forests and wetlands has greatly enhanced the production of particular types of timber, fibre, fish, and other products that are useful to people — but at the cost of substantially distorting the composition of the ecosystems concerned.

Table I (from Holdgate [1980]), based on a classical analysis by H.T. Odum [1967]), summarizes the gains in human food production, and in human populations supported, as a result of the transition from a hunter–gatherer life-style to industrialized modern agriculture. The examples cited, it must be emphasized, are not directly interchangeable: tropical rain-forests and semi-arid range-

TABLE I

ENERGY INPUTS, HUMAN FOOD, AND POPULATION SUPPORTED, FROM FOUR LEVELS OF DEVELOPMENT (ADAPTED FROM INSTITUTE OF ECOLOGY [TIE, 1971]).

System	Energy (kJ/m^2 year)		Human food (kJ/m^2 year)		Population supported	
	Solar	Industrial	Plant	Animal	On land	In city (km^2)
Rain-forest						
(hunter–gatherer)	6.3×10^6	0	1.7		1	0
Uganda grasslands	6.3×10^6	0	82	0.8	25	0
India:						
monsoon zone	6.3×10^6	0	1,029	113	230	30
United States:						
arable	6.3×10^6	567	4,200	0	60	2,000

lands are not generally suitable for conversion to intensive agriculture. Moreover, as yields are pushed up, maintenance costs and industrial inputs also rise (Fig. 1), so that sustainability demands a yield — and consequently a capacity to support people — that is short of the theoretically-attainable maximum. None the less, advances have been great: to cite one illustration of gains in a particular area, wheat yields per hectare in Britain increased tenfold or more between AD 1200 and 1970 (Cooke, 1970).

This enhancement of food production has been a crucial factor in supporting growing world populations (see earlier Chapters in this volume). Agricultural output and food production increased in both developed and developing countries in the period AD 1970–90 (FAO, 1988, 1991; cf. Fig. 2). Over most of that period in most parts of the world, the increase kept ahead of population growth, so that food production per head rose overall in both developed and developing countries between 1978 and 1989 (Fig. 3). However, this generally encouraging statistic overlies a deterioration that was severe in sub-Saharan Africa, where production per head in 1989 was only 92% of that in 1979–81. In the developed countries, the annual rate of

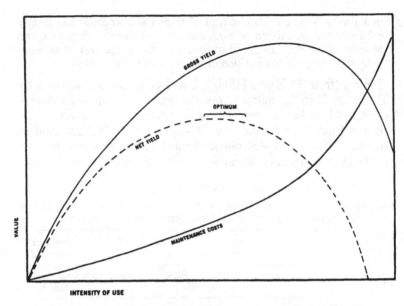

FIG. 1. *Trends in gross agricultural yield, maintenance costs, and net yield with increasing intensity of use (from TIE, 1971).*

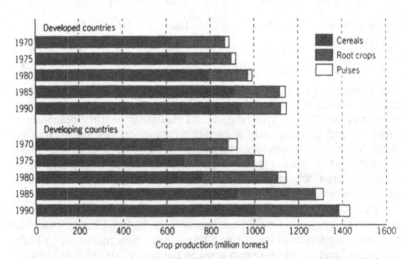

FIG. 2. *Crop Production in developed and developing countries, 1970–90 (from Tolba et al. [1992] based on FAO, 1988, 1991).*

increase in cereal production was about twice that of population growth, but in the developing countries the margin was much lower (about one-fifth as much) (Tolba *et al.*, 1992). Wide disparities between nations and peoples emerged as a consequence, as Chapter 9 indicates.

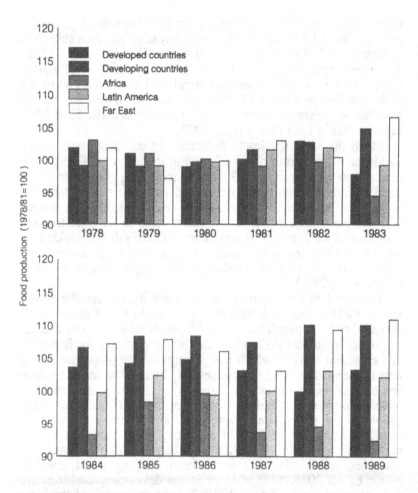

FIG. 3. *Indices of* per caput *food production, 1978–89, respectively for developed countries, developing countries, and three developing regions of the world (from Tolba et al., [1992] based on FAO, 1990).*

These increases, even in developed countries, were not achieved without cost. Fertilizer use has increased enormously, and soil erosion has become a widespread problem. World-wide, about 295 million hectares of land — an area only slightly less than that of India — has suffered severe degradation, about 38% of it as a result of deforestation, 25% through overgrazing, and 28% through bad agricultural management. A much larger area — 910 million hectares or 'three Indias' — is moderately degraded, and while it is still suitable for farming, has lost a considerable part of its productivity (Tolba *et al.*, 1992). An analysis in 1980 suggested that some 79% of the world's irrigated land, rain-fed cropland, and rangeland (pastures), was affected by processes leading towards degradation. Between 30 and 80% of all land under irrigation was then subject to salinization, alkalinization, or waterlogging (Holdgate *et al.*, 1982). The worst degradation is in Africa, followed by Asia — two regions that are least of all able to sacrifice irreplaceable resources on such a scale. Another problem is the increasing number of pest species that have become resistant to pesticides — recorded as under 50 in 1955, but over 500 in 1988 (Tolba *et al.*, 1992).

Much the same applies to fisheries. On one hand, world-wide yields have risen from about 67 million tonnes in 1970 to a little over 100 million tonnes in 1990, and an increasing proportion of this yield is coming from directly-managed aquaculture. On the other hand, over-fishing has led to substantial declines in size of a number of exploited stocks, especially in waters close to the industrialized countries of the Northern Hemisphere, with a result that yields now are lower than could be sustained even if fishing efforts were reduced, and the size and quality of fish caught has also declined (Tolba *et al.*, 1992).

The direct enhancement of carrying capacity, through the expansion in yield of food and other biological products that are useful to people, has been paralleled by modifications of the habitat and the use of technological skills to create enhanced shelter for humanity and enhanced services which now support a large proportion of the human population. These processes may broadly be termed urbanization and industrialization, respectively. Sophisticated shelters, which maintain constant near-optimum microclimates and have supplies of potable water, waste disposal services, and instantly-available energy in the form of electricity, are a feature of a great part of the world. In 1990, 72.6% of the population of developed countries, and 33.6% of those in the developing world, lived in cities: by AD 2025, 57% of those in the developing countries are expected to do so (Tolba *et al.*, 1992). Although many of the latter experience only the most basic conditions and suffer from 'the pollution of poverty' and an acute lack of essential services, provision of these services is slowly increasing.

Urban and industrial development have permitted an increasing number of people to engage in crafts and manufactures which have provided many goods and materials that are fed back into the enhancement of carrying capacity in agriculture and also raise people's quality of life in many other ways. The development of modern transport systems has meant that communities are no longer dependent on local produce, and it is possible, albeit at often unacceptable cost, to transport food surpluses from areas of over-production to areas of shortage. The development of transport indeed brought with it changes in the pattern of land-use over whole continents — including North America, where in the 19th century the eastern hills of New England were largely stripped of forest and subjected to mixed farming, only to become abandoned and reforested when agricultural productivity on the Great Plains rose and produced abundant cereals at much lower cost (Turner *et al.*, 1990).

THE IMPERATIVE FOR 'SUSTAINABLE DEVELOPMENT'

Numerous volumes and papers have described the other side of the developmental coin: soil erosion, land degradation and desert-ification, deforestation, losses of biological diversity and genetic resources, overexploitation of fisheries, pollution of soil, water, and air, stratospheric ozone depletion, and the threat of climatic change as a consequence of 'greenhouse gas' emissions especially from the energy-supply industries of the developed world. They are well summarized in the two major reports by UNEP on the State of the Environment (Holdgate *et al.*, 1982; Tolba *et al.*, 1992), in the Pro-ceedings of four International Conferences on Environmental Future (Polunin, 1972, 1980; Polunin & Burnett, 1990, 1993), and in the review *The Earth as Transformed by Human Action* (Turner *et al.*, 1990). These issues are important: first, because they impede and undermine the development process, erode 'carrying capacity', and aggravate the problems for increasing human populations, and, secondly, because the recognition of their severity has, especially latterly, been a major spur to political and social action.

The United Nations Conference on the Human Environment, held at Stockholm in 1972, and the United Nations Conference on Environment and Development, held in Rio de Janeiro 20 years later, were manifestations of global concern over such problems. They were also landmarks in our recognition that, while development is essential, the nature of the development process must change. Development, too, has to work within the limits of the Biosphere, and recognize the complexity and sensitivity of its processes. Only if development is sustainable, will it succeed in bringing millions of people out of acute poverty and cater for the five thousand millions or so of further individuals which demographic projections suggest will be added to the world community in the next half-century.

Agenda 21, the international action plan adopted by the unpre-cedentedly large gathering of Heads of State and Government at UNCED in Rio de Janeiro (Robinson, 1993), emphasizes the need for fundamental changes in consumption and resource-use patterns, and substantial reductions in the release of pollutants. The fact that this need has been recognized is a sign of hope. Even more important is the circumstance that many lessons have now been learnt about the nature of the development process, including the following:

1) sustainable development depends on the way in which humanity cares for the wider ecological processes of the Biosphere, avoiding disruptive impact on climate, the distortion of biogeochemical cycles, the attenuation of the stratospheric ozone layer that shields ecological systems and humanity from damaging solar ultraviolet B radiation, and the attrition of genetic resources (IUCN/UNEP/WWF, 1991). 'Wild Nature' is surely as important to the human future as is that sector of the Biosphere which has been brought under human management;

2) it is no longer acceptable to treat these natural processes, which constitute the human life-support system, as 'free goods' — *i.e.* outside the economic equation, and hence very largely outside the political decision-taking process and the operation of the market system. New systems of economic valuation which cover these crucial parameters are urgently needed (McNeely, 1988). Such valuation systems, and new economic models, must also incorporate the other economic benefits that natural environments provide to humanity — for example through the regulation of hydrological cycles, protection of coasts from erosion, nurturing of fishes in coral reefs and mangroves, and maintenance of biological diversity;

3) the continuing destruction of the world's biological diversity, largely as a result of conversion of diverse natural habitats — especially tropical rain-forests — into relatively species-poor cultivated ground and range-land, brings with it substantial risks. This is partly because agriculturalists themselves depend on the wild relatives of cultivars for genetic material to incorporate into cultivated strains — to give the latter enhanced resistance to pests and diseases and also tolerance of climatic variations. It is also because it is now recognized that the genetic resources in wild systems have immense value to humanity as a source of medicinal and other products which have not been identified and valued at all adequately by humanity over the millennia of selective domestication of crop and other species;

4) human use of the diverse resources of Nature is selective, traditionalist, and inefficient. Only a handful of plant and animal species have been cultivated or domesticated. Some natural ecosystems and wild species, cropped sustainably, can in fact provide a yield that is economically and socially preferable to traditional forms of land-use. The German and British cultural and technical traditions of managing forests predominantly as sources of timber, harvested by clear-felling, is being modified on due recognition that tropical forests can be cropped continuously and sustainably to provide a vast range of products — including food, medicines, fibres, resins, and oils, as well as carefully-extracted timber — with an *annual* revenue per hectare which can be greater than that derivable only once through the cutting of the forest and sale of the timber (Myers, 1988; Holdgate, 1993; *see* Fig. 4). In a similar way, cropping wild animals grazing native range can be economically more productive than replacing them by cattle, which

often need to be sustained by modified herbage that is more prone to erosion and diseases than the natural vegetation. This is the basis of the CAMPFIRE project in Zimbabwe, and of course of the ecotourism that is the principal foreign-exchange-earner in a number of developing countries;

5) the assumption that population growth leads inexorably to poverty and environmental degradation is false. Sound processes of rural development, especially based on leadership within, and empowerment of, the rural communities themselves, can achieve improved environmental quality, reduce poverty, and enhance carrying capacity, despite the growth of human numbers in the areas concerned. The key is in the nature of the environment and the development process, and the close and effective involvement of the people; and

6) it is also fallacious to assume that the problems of the world are confined to the developing countries, where pressure on resources, poverty, and population growth, are visibly most acute. Resource use in the developed countries is also often far beyond sustainable limits. Moreover, these countries consume and waste resources excessively. The developed world, with about a quarter of the world's population, was accustomed in 1992 to consume some 70% of total commercial energy, 75% of traded metals, 85% of marketed wood, and 60% of food (Tolba et al., 1992). That same c. 25% of the global population living in the developed and urbanized regions also accounted for more than half the pollution which is threatening to perturb global climate, degrade the stratospheric ozone screen, and impair the fertility of ecosystems through the long-range transport and deposition of pollutants such as those in 'acid rain'. If the under-industrialized three-quarters of the world is to industrialize — and this seems likely to be an important condition for achieving population stability — then the developed countries will need to reduce their overconsumption of energy and materials and their over-production of pollutants. The quest for global sustainability affects all nations.

Sustainable development of the right kind, i.e. which pays due attention to environmental/ecological factors and heeds the lessons learned of late, can certainly enhance the carrying capacity of the Earth and make it possible to support an increased number of people with a higher quality of life. It must be emphasized, however, that this will only be possible if both developed and developing countries change many of their current practices and assumptions. An obvious imperative is to halt the loss of food-productive land. It has been estimated that between AD 1975 and the year 2000, about 200 million hectares of agricultural land will be transformed into non-agricultural land, that another 50 million hectares will be seriously degraded, and that yet another 50 million hectares will be desertified (Buringh & Dudal, 1987). These losses have to be compensated for either by raising yields or by cultivating land that is not currently under agricultural use, simply to 'stay where we are now'. And to

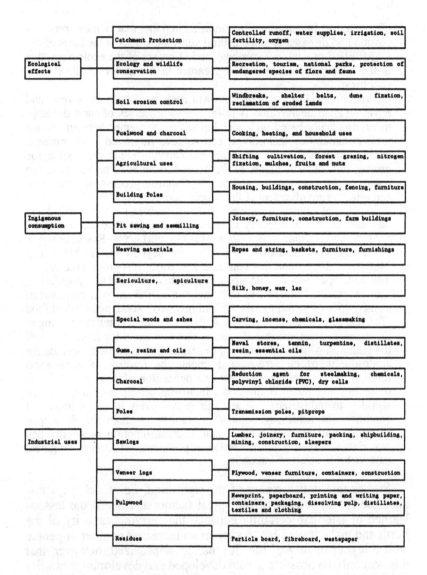

FIG. 4. *Spectrum of goods and services available from tropical moist forests (from Myers, 1988).*

meet the needs of the people who will be added to the world population in that period, a further 300 million hectares (or an equivalent increase in yield from existing farmland) will be needed. The world cannot afford the waste that is represented by mismanagement, by the encroachment of cities onto the more fertile lowlands, and by pollution.

There are limits, even if uncertain ones, to the extent to which carrying capacity can be raised. It has been calculated that humanity currently appropriates or destroys some 39% of the primary productivity of green plants on land, and 2.2% of aquatic primary production (Vitousek *et al.*, 1986). There are theoretical grounds for doubting whether humanity can ever hope to use a much higher percentage of global primary productivity, even if biotechnology dramatically increases the productive efficiency of particular cultivated species or varieties. What we can do is cut the waste back — and the means for doing this by improved land management, control of urban and industrial expansion, and strict pollution control, exist. Technology should not be seen as a means of stretching planetary carrying-capacity indefinitely. What is needed is that these skills be used to win time within which the demographic transition can be accomplished and a stable human population achieved within the limits of a Biosphere that remains ecologically diverse, retains its functional integrity, and provides for resource uses by humanity that are ecologically sustainable. The lower the figure at which global human population stabilizes, the higher the probability of success will be (Holdgate, 1996).

SECURING THE DEMOGRAPHIC TRANSITION

There are five main factors which need to interlink in order to secure human population stability, namely:

1) the elevation of local carrying-capacity to an optimum level that is sustainable, with the optimum mix of natural-resource use;
2) the curbing of pollutant and other impacts from wider regions which could undermine that local carrying-capacity;
3) the provision of medical care, including the means for birth-control and measures for minimizing infant mortality;
4) economic development that provides a prospect for employment and a reasonable quality of life for all people; and
5) educational, social, and cultural, support, informing people about the wisdom of reducing family size to sustainable levels, and influencing beliefs and outlooks to that end.

The first two of these have already been reviewed. The important point is that all five are interlinked and part of one socio-cultural context. This is increasingly evident from the patterns of population changes which we see around us, and is increasingly recognized in an emerging global consensus. Both the historical demographic changes and the global policy changes offer hopes for the future, but they also point to imbalances and problems that must be addressed.

Let us first examine the historical demographic changes, and then the emerging global consensus as seen in a series of recent international conferences.

The demographic changes can best be summarized under the rubric of the demographic transition, that remarkable change of the past two centuries in which populations have moved from high to low birth- and death-rates. Progressing from high to low birth- and death-rates has also meant progressing from relatively poor, rural-agrarian society to the modern urban industrial society that offers a higher quality of life, but also threatens to be unsustainable. The important point of the demographic transition for our present discussion is that there is not one transition, but actually two — namely a past and a present.

Figure 5 shows the demographic transition in England and Wales, which we use to illustrate the past. Death-rates began to decline gradually around AD 1700, while the birth-rate remained high and even increased very slightly. This gave rise to a period of relatively rapid population growth, from less than 0.5% to just over 1.0% per year. From the middle of the 19th century, the birth-rate began to decline, finally coming practically into line with lowered mortality to produce once again a period of slow growth (*cf.* Fig. 5). All currently industrialized societies of the world have gone through this transition, though they differed somewhat in the trajectories, for the process was rather gradual, and until relatively recently was not accompanied by major new developments in medical or contraceptive technology.

Having gone through this demographic transition, all advanced industrial societies now show total fertility rates at less than replacement level.* Some are projected to show negative growth-rates early in the next century. All are becoming concerned about the increasing proportion of elderly people in their populations — especially given the increasing skill-demands and costs of geriatric care — and the proportionate shrinkage of the working-aged population of ages 20–65. But in global and national terms, the fact that the population of the more developed regions† is projected to remain effectively stable over the next several decades, is a sign of

---

* The total fertility rate (TFR) is, roughly, the number of children a woman will bear through her reproductive years. In 'traditional' societies it is usually at the level of 6 to 8. When the TFR falls to 2.1, it is at replacement level. If a society maintained a TFR of 2.1 for some years, births and deaths would be equal, and the population numbers would be stationary, at least in the absence of migration.

† This is a term used by the United Nations Population Division, which classifies all countries as More Developed or Less Developed, with a sub-category of the latter to include the 40 Least Developed. The More Developed include the industrialized countries of Europe and North America, together with Australia, New Zealand, and Japan.

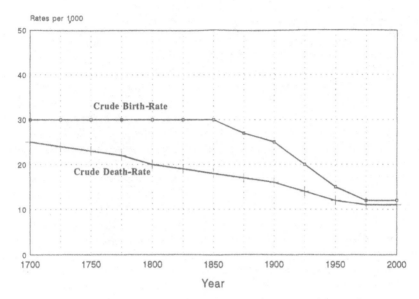

FIG. 5. *Past Demographic Transition in England and Wales, 1700–2000. Source: UN, 1994.*

hope. For the developed world, the challenge is now to achieve economic and social adjustment within communities, with a diminished proportion in the actively working population, and major reductions in resource and energy consumption and concomitant pollution emissions.

The second demographic transition is now evident throughout the Less-developed regions, as illustrated in Figure 6. It is moving quite rapidly, but is at different stages in different regions (UN, 1994). In some (China, Hong Kong, Mauritius, South Korea, Singapore, Taiwan, and Thailand) it has been completed, and fertility is now at or below replacement level (cf. Fig. 7). In other countries mortality has fallen rapidly, but fertility, though declining, remains above replacement level. This is the case in some of the states of India, Indonesia, Malaysia, Tunisia, and most of Latin America. In yet other countries, mortality has fallen but fertility remains high, resulting in rates of population growth that are near 3% per year. This is the case in Pakistan, the Middle East, and throughout most of Africa.

The different speeds of the two demographic transitions have important implications for global security. The past demographic transition took place more gradually from relatively lower levels of mortality and fertility. Thus the intervening period of rapid growth implied rates of about 1% per year. The current, second demographic, transition started from much higher levels of both mortality

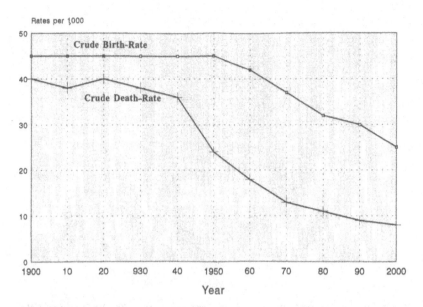

FIG. 6. *Present Demographic Transition in Africa, Asia, and Latin America. Source: UN, 1994.*

and fertility, and the mortality decline has come much more rapidly — so that current growth-rates of 3% per year, or more, have characterized many countries prior to the onset of fertility decline. It is also noteworthy that these differences have profoundly different causes, both in terms of technology and of policy (which in turn relates closely to culture).

The declines of fertility and mortality in the industrialized countries took place against a background of relatively slow advances in medical skill and very little contraceptive technology. Mortality declined with slow advances in the standard of living, and with the introduction of public health technology, especially for urban water-supply and waste disposal. Major factors in fertility reduction were the reduced economic value of children and the improving status of women. The increasing range of opportunities provided for women within urban industrial societies, the spread of universal education, and a rise in the age of marriage, were all closely interwined, and reflected an increase in the status and autonomy of women — all of which supported a reduction in fertility.

In contrast, the demographic transition that is taking place today in the less-developed regions can show much more rapid change than took place in the industrialized countries. Modern medical technology has achieved some remarkable breakthroughs, which can produce mortality declines in a few years in contrast to the decades or centuries that were required in the past even in developed countries. Insecticides and anti-malaria campaigns launched by the WHO cut infant mortality rates drastically in just a few years. Inoculations and antibiotics to control infectious diseases brought rapid declines in mortality. Smallpox has been eliminated,* and malaria greatly controlled, though now it is making a comeback. Improvements in providing clean drinking-water and safe waste-disposal added to the rapid decline in mortality.

Similar advances have been made in contraceptive technology, and it is instructive to recall how recent these are. Oral contraceptive pills and intrauterine contraceptive devices have been widely available only since about 1965. In addition, we now have safer methods for sterilization, using injectables, sub-dermal implants, and a great increase in effective barrier methods for fertility limitation. Like the new mortality-controlling technology, these can be distributed easily through primary health-care systems or even through the market-place. The diffusion of the new contraceptive technology can move very rapidly, as China and Thailand show in Figure 7.

Both of these technological breakthroughs (in mortality and fertility control) offer great promise, but that promise is not realized everywhere, reflecting the imbalances of social services which we find throughout the less-developed regions. Primary health-care is still not available to massive numbers of the poor. Clean drinking-water and waste-disposal lag far behind the demand, especially in poor countries and among the poor in more wealthy countries. The same is true of contraceptive technology. As Figure 10 shows (UNFPA, 1993), in many countries women are having more children than they want, as they still do not have access to safe and effective modern contraceptives. *Correcting these imbalances constitutes one of the great challenges of our time.*

It is now widely recognized that, to achieve the full demographic transition, with declines in both mortality and fertility, health and fertility policies must be carried out within a much wider socio-economic context than before. This must include universal education, improved social infrastructure, cleaner environments, improved medical services, greater attention to women's health and welfare, and greater opportunities for women. The kind of expansion we have seen in school education projects (*cf.* Figure 9 [UNFPA, 1993]) is hopeful. Where health services and education are made more fully available, especially for women, we find that both mortality and fertility can decline rapidly (Caldwell, 1986).

In policy as well as technology, the current demographic transition has been associated with changes that are truly revolutionary. Prior to 1952, most governments had policies that encouraged high birth-rates and population growth. Throughout the history of govern-

---

* *See* footnote on p. 156.

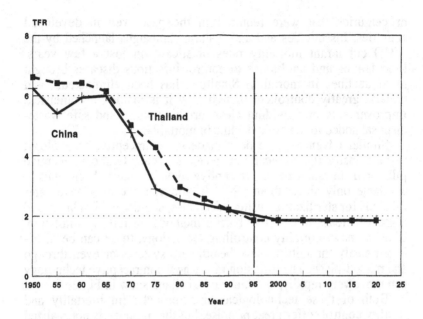

FIG. 7. *Total Fertility Rates for China and Thailand. Source: Ness, 1994. The stable population shown after 1995 is a more projection.*

ments, people have been seen as a source of power — a resource that can be taxed, worked, and sent to war. Thus governments have typically wanted more people and more births. In the 1930s countries such as France, Germany, Italy, Japan, and the United Kingdom, pursued such pro-natalist policies. In Japan and the United States, advocates of birth-control were imprisoned in the 1930s. Thus the fertility declines of the past demographic transition came about often against the policies of governments. The contrast with today could not be more striking, or gratifying if our descendents are to have a decent future.

In 1952 India became the first country to adopt an official policy to limit population-growth by limiting fertility within marriage (Ness & Ando, 1984).* Pakistan followed in 1960 and today such policies are quite common. More than 90 percent of the population of the less developed regions lives under governments that actively encourage fertility limitation, or provide support to private family

* Japan also adopted policies to permit birth-control and abortion from 1948 to 1952, but more as a health measure than to slow population growth. The great social and economic pressures following World War II led to a large increase in abortions, which were often unsafe and caused high rates of maternal mortality. In the absence of an effective contraceptive technology, the government made abortion available largely to improve the health of women, which it did quite remarkably (Ness & Ando, 1984).

planning associations. The fertility declines in the current demographic transition are therefore taking place with the strong support of national governments.

Government population policies, in the manner of those for social services in general, vary considerably in their strength and effectiveness. In some countries — such as China[†], Indonesia, Korea, Mexico, Sri Lanka, and Thailand — there are strong policies for population control. In India, Bangladesh, Tunisia, Botswana, Kenya, Zimbabwe, and most of Latin America, there are positive policies, but of a more moderate nature (cf. Figure 8). In Pakistan, Nigeria, and Ghana, policies are quite weak. On the other hand, there are some countries — including Burma and Saudi Arabia — where governments remain strongly pro-natalist. While there has been a general movement towards stronger and more effective population-control programmes, some countries have swung back and forth from pro-natalism to anti-natalism.

While these transitions have been taking place, there has been a hopeful associated set of changes in the stated policies of the global community. Following the first International Conference on Environmental Future in 1971 (Polunin, 1972), the first United Nations' Conference on the Human Environment was held in Stockholm in 1972, followed 20 years later by the United Nations Conference on Environment and Development (UNCED, cf. Nazim & Polunin, 1993). There have now been three international population conferences, in Bucharest 1974, Mexico City 1984, and Cairo 1994.[§] These five United Nations' conferences and the four International Conferences on Environmental Future represent a growing global consensus on the necessity to reduce population growth to produce a higher quality of life for all humanity, in both present and future generations, than would otherwise be foreseeable. Moreover, they reflect a growing consensus that the problem of sustainability is not simply one of reducing population growth in the less-developed countries, but equally requires a change in patterns of production and consumption in the more-developed regions.

[†] China's fertility limitation policy is often criticized for its coercive aspects. While there is some merit to the criticisms (which are even made by the Chinese authorities themselves), it is important to note that the birth-control policy came 20 years after the radical primary health program, in which China mobilized more than a million 'barefoot' doctors to provide basic medical care, especially controlling infectious diseases, throughout most of the countryside. Without the development of this massive primary health-care delivery system, it is doubtful that a coercive policy of birth-control could have been at all successful.

[§] The first two of these conferences were officially called the International Conference on Population. The Cairo conference in 1994 was officially renamed, by the UN General Assembly, the International Conference on Population and Development (ICPD).

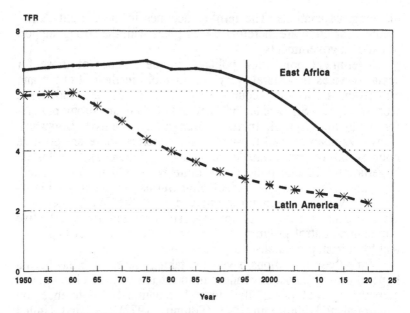

FIG. 8. *Total Fertility Rates for East Africa and Latin America.* Source: Ness, 1994.

It is often stated that the population issues were evaded in the United Nations Conference on the Environment and Development (UNCED), held in Rio de Janeiro in June 1992, and that this was because of the cultural and religious sensitivity of the subject.* This is only a partial truth. Population issues are prominent in many of the individual countries' policy papers prepared for the conference. While they occupy only 18 pages, population issues are discussed in Chapter 5 of *Agenda 21,* entitled, 'Demographic Dynamics and Sustainability' (Robinson, 1993). That chapter correctly emphasizes that demographic trends and factors and sustainable development have a synergistic relationship. It states (para 5.17) that 'full integration of population concerns into national planning, policy and decision-making processes should continue. Population policies and programs should be considered with full recognition of women's rights'. In para 5.43, *Agenda 21* emphasizes that 'population programs should be implemented along with natural resource management and development programs at local level, that will ensure sustainable use of natural resources, improve the quality of life of people, and enhance environmental quality.'

---

* This reminds us of the warning we remember being given before our first International Conference on Environmental Future in 1971, namely that we should keep off the subject of human population because it was rumoured that a strong Brazilian delegation was expected. — Ed.

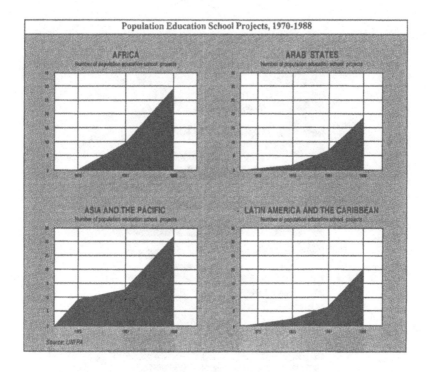

FIG. 9. *Population Education School Projects 1970–1988 (from UNFPA, 1993).*

Furthermore, although birth-control programmes are not pressed urgently (and this has been the cause of much of the criticism), paragraphs 5.49 and 5.50 of *Agenda 21* do state that 'reproductive health programs and services should, as appropriate, be developed and enhanced. Government should take active steps to implement, as a matter of urgency, in accordance with country-specific conditions and legal systems, measures to ensure that women and men have the same right to decide freely and responsibly on the number and spacing of children...'. Such statements correctly recognize that population stability can be attained only through a broader process which blends the control of fertility with the social and economic development process. This is fully in accord with the empirical evidence summarized and the approach advocated in the present Chapter.

The three UN international population conferences cited above have seen a steady rise in international consensus on the need for population policies to be set in a broader socio-economic context. The 1994 Cairo ICPD was especially remarkable on a number of accounts. Least important, but perhaps most memorable, was the

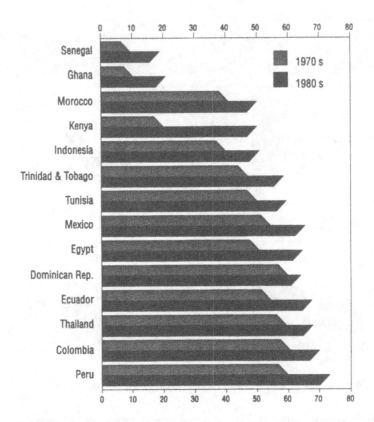

FIG. 10. *Percentage of Women Wanting No More Children — increasing in all 14 cited countries in the 1980s over 1970s, and in Kenya more than doubling (from UNFPA, 1993). Source: Demographic and Health Surveys, various years.*

attention of the media which was brought about by the unusually strong Vatican opposition to the Conference itself and to some of the language of the programme of action. This assured constant media attention, a flood of reporters and television cameras, and to some extent detracted attention from the extensive technical issues that were being raised. Ultimately, however, the ICPD can be counted as a substantial step forward, providing more grounds for hope.

The Vatican opposition did focus attention on the ethical, moral, and religious, issues that are of profound importance in population policy. Moreover, on the last day of the Conference, the Vatican announced that it wished to 'join the consensus' of the overall programme of action, because '... it opens out some new paths concerning the future population policy,' said Archbishop Renato Martino, the Vatican delegate. Along with some other countries, it withheld acceptance of certain provisions, but this was, nonetheless, the first time the Vatican had joined the consensus.

The language of the World Programme of Action (UN, 1995) adopted at Cairo is rich in explicit reference to a wide range of issues of development, human welfare, and environmental protection. From the 15 principles on which the Programme is based, the following excerpts reflect some of this broad concern:

> Principle 2. Human beings are at the centre of concerns for sustainable development. They are entitled to a healthy and productive life in harmony with nature...

> Principle 3. ... The right to development must be fulfilled so as to meet equitably the population, development, and environmental, needs of present and future generations.

> Principle 6. Sustainable development as a means to ensure human well-being, equitably shared by all people today and in the future, requires that the interrelationships between population, resources, the environment, and development, should be fully recognized, properly managed, and brought into harmonious, dynamic balance. To achieve sustainable development and a higher quality of life for all people, States should reduce and eliminate unsustainable patterns of production and consumption and promote appropriate policies, including population-related policies, in order to meet the needs of current generations without compromising the ability of future generations to meet their own needs.

Chapters III through XI of the Programme provide a wealth of technical detail on the full range of linkages between population, development, and the environment. Much of this content was developed in a series of preparatory committee meetings before the Conference, in which the great majority of items were agreed upon. The technical points deal with the relation between population, development, and environmental protection (Chapter III), Gender Equality and the Empowerment of Women (Ch. IV), Population Growth and Structure (Ch. VI), Health, Morbidity, and Mortality (Ch. VIII), Urbanization (Ch. IX), International Migration (Ch. X), and Education. Chapter VII, on Reproductive Rights and Reproductive Health, posed the most serious problem, because of its inclusion of issues on contraceptives and safe abortions, but these were only a small fraction of the pertinent issues that received extensive attention at Cairo.

Controversies remain, and will no doubt exercise many people in the near future. For example, language on reproductive health that was easily accepted in 1992 at the UNCED in Rio de Janeiro, was subject to question at Cairo, and became increasingly controversial in the context of the World Summit on Women, held in Beijing in 1995. Nonetheless a consensus was achieved at Cairo and a programme of action was laid down that fits well with the evidence of population change discussed above. It is a consensus for a broadly-based set of activities to address the population issue, taking a position with which this Chapter is fully in accord. Let us end our figures with two of the most encouraging ones already cited in earlier contents, namely Fig. 9 showing the steady rise in 'Population Education School Projects', starting in 1970, and Fig. 10 showing the 'Percentage of Women Wanting No More Children' — increasing in all the 14 cited countries in the 1980s over 1970s'.

SIGNS OF HOPE

There are, therefore, many signs of hope all around us. They include in particular:
1) the near-universal recognition of the need for really sustainable development, based on the understanding of environmental carrying capacity and the management of environmental resources to stay within ultimate limits, avoid environmental degradation, and optimize the yield and use of the products which a particular area can best deliver;
2) recognition that population increase does not necessarily and automatically bring environmental degradation: sensitively managed, it is perfectly possible to sustain an increasing number of people in an area and even improve its environment, provided that this is done using patterns of management and development that are indeed sustainable;
3) increasing acceptance over most of the world that it is imperative to bring humanity into balance with the environment, and that this is an essential component of the development process; and
4) recognition that this progress is only attainable within overall development strategies which create the appropriate social and medical infrastructure, provide education, recognize the important role of women as well as men in a modern society, and yet are culturally sensitive, promoting population regulation policies within an acceptable context.

The most obvious sign of hope is actually in what is happening in the world today. The populations in most of the industrialized world are at or near stability. A number of rapidly industrializing countries, and some that are still at an earlier stage in development, have gone through the demographic transition towards population stability. (This is the case in much of South-east Asia.) The (second) demographic transition is proceeding much more rapidly in the developing world as a whole than it did in the industrialized countries that went through the first demographic transition. Thus although birth-rates are still extremely high in a number of countries, in general they are falling fast, and the process that took countries such as England and Wales 250 years to achieve, seems likely to take the developing world less than a century.

It is remarkable that the process of demographic transition which is now so widely visible in the world, does not seem to be culture-specific, despite the accusations that are commonly levelled against particular religions and cultures as obstructions. In Latin America, fertility has been declining steadily, and the demographic transition in South and Central America as a whole will, on current projections, be complete by around the year 2025. Some Islamic countries have also been following very strong policies towards population regulation.

There is an English verse which begins 'If hopes were dupes, fears may be liars'. Extrapolation of trends clearly provides an only partially reliable basis for predicting the future; but if present trends continue, then the worst fears of population crashes, global environmental degradation, and escalating poverty and strife, may prove to be liars. But equally, hopes can still be dupes. The key to making hopes real, resides only partly in the continued effort to promote social policies that lead people to regulate the numbers of their children. It lies also in the wisdom and efficiency with which they use the finite resources of the Biosphere.

If humanity applies the understanding and knowledge which it now has, there is good prospect that an increasing proportion of the world will achieve stable populations within environments that are managed sustainably and to a higher quality than today, with global pollution curbed and over-use of resources substantially cut back by the middle years of the coming century. That must be the goal of the world's governments as they implement *Agenda 21*.

ACKNOWLEDGEMENT

Our grateful thanks are due to UNEP and UNFPA for permision to reproduce diagrams.

REFERENCES

Agenda 21 — See ROBINSON, N.A. (1993).

BURINGH, P. & DUDAL, R. (1987). Agricultural land use in space and time. Pp. 9–43 in *Land Transformations in Agriculture* (Eds M.G. WOLMAN & F. FOURNIER). (SCOPE 32.) John Wiley & Sons, Chichester, England, UK: xix + 531 pp., illustr.

CALDWELL, J. (1986). Routes to low fertility in poor countries. *Population Development Review*, **12**(2), pp. 171–220.

COOKE, G.W. (1970). The carrying capacity of the land in the year 2000. Pp. 15–42 in *The Optimum Population for Britain* (Ed. L. R. TAYLOR). Institute of Biology Symposium No. 19, Academic Press (for the Institute of Biology), London & New York: 183 pp., figs and tables.

296   MARTIN W. HOLDGATE & GAYL D. NESS

CROSBY, A.W. (1986). *Ecological Imperialism: The Biological Expansion of Europe, 900–1900.* Cambridge University Press, Cambridge, New York, Port Chester, Melbourne & Sydney: 7 figs, 20 plates. [Not available for checking.]

FAO (1988). *Country Tables.* Food and Agriculture Organization, Rome, Italy: [not available for checking: cited from Tolba *et al.* (1992).].

FAO (1990). *FAO Production Yearbook*, Vol. 42. Food and Agriculture Organization, Rome, Italy: v + 350 pp., tables.

FAO (1991). *The State of Food and Agriculture — 1990.* Food and Agriculture Organization, Rome, Italy: xxii + 223 pp., tables.

HOLDGATE, M.W. (1980). Man's impact on environmental systems. Pp. 187–209 in *Food Chains and Human Nutrition* (Ed. K. BAXTER). Applied Science Publishers, Barking, Essex, England, UK: x + 470 pp., figs and tables.

HOLDGATE, M.W. (1993). *Sustainability in the Forest.* (Keynote Address to the 14th Commonwealth Forestry Conference, Kuala Lumpur, Malaysia, 13–18 September 1993.) *Commonwealth Forestry Review*, **72**(4), Nr 231, pp. 217–25, fig. and 4 tables.

HOLDGATE, M.W. (1996). *From Care to Action.* Earthscan Publications Ltd, 120 Pentonville Road, London N1 9JN, England, UK: xviii + 346 pp., figs & tables.

HOLDGATE, M.W., KASSAS, M. & WHITE, G.F. (Eds) (1982). *The World Environment, 1972–1982: A Report by the United Nations Environment Programme.* Tycooly International (for the United Nations Environment Programme), Dublin, Ireland: pp. xxxii + 637, figs and tables.

HUTCHISON, R.A. (1991). *Fighting for Survival: Insecurity, People and the Environment in the Horn of Africa.* International Union for Conservation of Nature and Natural Resources, Gland, Switzerland: 181 pp., 18 figs.

INSTITUTE OF ECOLOGY — *see* TIE (1971).

IUCN/UNEP/WWF (1991). *Caring for the Earth: A Strategy for Sustainable Living.* Switzerland, published in partnership between IUCN — The World Conservation Union, UNEP — The United Nations Environment Programme, and WWF — World Wide Fund for Nature, Gland, Switzerland: 228 pp., figs, tables, and 31 boxes.

MAGUIRE, A. & BROWN, J.W. (Eds) (1986). *Bordering on Trouble: Resources and Politics in Latin America.* Adler & Adler (for World Resources Institute), Bethesda, Maryland, USA: 447 pp., tables, maps.

MCNEELY, J.A. (1988). *Economics and Biological Diversity: Developing and Using Economic Incentives to Conserve Biological Resources.* International Union for Conservation of Nature and Natural Resources, Gland, Switzerland: 236 pp.

MYERS, N. (1988). Tropical forests: much more than stocks of wood. *J. Tropical Ecology*, **4**, pp. 209–21, figs.

NAZIM, MOHAMMAD & POLUNIN, N. (Eds). (1993). *Environmental Challenges: From Stockholm to Rio and Beyond.* Energy and Environment Society of Pakistan, Lahore, Pakistan & Foundation for Environmental Conservation, Geneva, Switzerland: vi + 284 pp., illustr. Foreword by Gro Harlem Brundtland.

NESS, G.D. (1994). *Population and the Environment: Frameworks for Analysis.* Working Paper Nr 10, The Environmental and Natural Resources Policy and Training Project (EPAT/MUCIA), University of Wisconsin, Madison, Wisconsin 53705-2397, USA: [vii] + 36 pp., 5 appendixes.

NESS, G.D. (1994). The Long View: population – environment dynamics in historical perspective. Pp. 33–56 in *Population Environment Dynamics: Ideas and Observations* (Eds G.D. NESS, W.D. DRAKE & S.R. BRECHIN). University of Michigan Press, Ann Arbor, Michigan, USA: ix + 456 pp., figs and tables.

NESS, G.D. (MS). *Notes on the Demographic Transition.* Available from Professor Gayl D. Ness, c/o IUCN/UICN, 28 Rue Mauverney, 1196 Gland, Switzerland: 4 pp. + charts.

NESS, G.D. & ANDO, H. (1984). *The Land is Shrinking: Population Planning in Asia.* Johns Hopkins University Press, Baltimore, Maryland, USA: xxii + 225 pp.

ODUM, H.T. (1967). Energetics of world food production. Pp. 55–95 in *The World Food Problem*, vol. 3 [not available for checking].

POLUNIN, N. (1972). *The Environmental Future: Proceedings of the First International Conference on Environmental Future, held in Finland from 27 June to 3 July 1971.* Macmillan, London & Basingstoke, England, UK: xiv + 660 pp., frontispiece, figures, and tables.

POLUNIN, N. (1980). *Growth Without Ecodisasters?: Proceedings of the Second International Conference on Environmental Future (2nd ICEF), held in Reykjavik, Iceland, 5–11 June 1977.* Macmillan, London & Basingstoke, England, UK: xxvi + 675 pp., frontispiece, figures, and tables.

POLUNIN, N. & BURNETT, SIR J.H. (Eds) (1990). *Maintenance of The Biosphere: Proceedings of the Third International Conference on Environmental Future (3rd ICEF).* Edinburgh University Press (for the Foundation for Environmental Conservation), Edinburgh, Scotland, UK: xvi + 228 pp., frontispiece, photographs, and diagrams.

POLUNIN, N. & BURNETT, SIR J.H. (Eds) (1993). *Surviving With The Biosphere: Proceedings of the Fourth International Conference on Environmental Future (4th ICEF), held in Budapest, Hungary, during 22–27 April 1990.* Edinburgh University Press (for the Foundation for Environmental Conservation), Edinburgh, Scotland, UK: xxii + 572 pp., frontispiece, diagrams, maps, and tables.

ROBINSON, N.A. (Ed.) (1993). *Agenda 21: Earth's Action Plan, Annotated.* IUCN Environmental Policy and Law Paper No. 27, Oceana Publications Inc., New York, London & Rome: xciv + 683 pp.

SMITH, C. WOODHAM (1962). *The Great Hunger, Ireland 1845–9.* Harper & Row, New York, NY, USA: 510 pp., illustr.

TIE (1971). *Man in the Living Environment: The Institute of Ecology Report on the Workshop on Global Ecological Problems.* University of Wisconsin, Madison, Wisconsin, USA: 267 pp., diagrams & tables.

TOLBA, M.K., EL-KHOLY, O., EL-HINNAWI, E., HOLDGATE, M.W., MCMICHAEL, D.F. & MUNN, R.E. (Eds) (1992). *The World Environment, 1971–1992: Two Decades of Challenge.* Chapman & Hall (on behalf of the United Nations Environment Programme), London, Glasgow & New York: xi + 884 pp., numerous figs and tables.

TURNER, B.L., CLARK, W.C., KATES, R.W., RICHARDS, J.F., MATHEWS, J.T. & MEYER, W.B. (Eds) (1990). *The Earth as Transformed by Human Action: Global and Regional Changes in the Biosphere Over the Past 300 Years.* Cambridge University Press (with Clark University), Cambridge, Port Chester, Melbourne & Sydney: xiv + 713 pp., photos, diagrams, maps & tables.

UN (1989). *World Population Prospects 1988.* United Nations, New York & Geneva: xii + 579 pp., illustr.

UN (1994). *World Population Prospects, the 1994 Revision,* Appendix Tables. The United Nations, New York, NY, USA: 255 pp.

UN (1995). *Population and Development: Program of Action adopted at the International Conference on Population and Development, Cairo, 5–13 September 1994.* (Preliminary version.) The United Nations, New York, NY, USA: 155 pp.

UNFPA (1993). *Population Issues: Briefing Kit, 1993.* United Nations Population Fund, NY, USA: 21 pp., diagrams.

VITOUSEK, P.M., EHRLICH, P.R., EHRLICH. A.H. & MATSON, P.A. (1986). Human appropriation of the products of photosynthesis. *BioScience,* **36**(6), pp. 368–73, illustr.

WOODHAM SMITH, C. — *See* SMITH, C. WOODHAM (1962).

INDEX (including ACRONYMS etc.)*

by

LYNN M. CURME

*Some familiar items which are apt to be cited practically throughout the book, may be indexed only the first time or two (followed by '...') they are mentioned in the text. This should at least establish the identity and clarify the meaning in which these items are employed, while warning meticulous scholars that there may be further pertinent citations of them in the text. — Ed.

ecologically-sustainable development in
  Asia 114
economic activity, world 123
– advancement at environmental cost 27
– aspects of child-bearing 60–1
– conflicts 1
– constraint to sustainability 244–7
– development 93, 140
– – and rising prosperity 84
– effects of depletion of natural capital
  258
– imbalances, increase of 55
– inefficiencies 27
– problems & obstacles 6
– refugees 176
– revaluation needed 280
– status 13
– sustainability 118–9
– symptoms of unsustainability 266
– theories of family fertility and family
  size 101
– weakening — migration 177
– well-being 93
economies, collapsed 258
–, world, regional, and national 174
economists — human population-levels
  94
economy, global 264
ecosystem 19
– services 29
ecosystems, forest 8
–, scarce 8
–, tropical 8, 9
education 13, 33, 103
education — nutrition 166–7
–, basic 38
– — fertility decline 197
–, increased, for women 236
–, maternal 167
educational opportunities 122
EEC 85
Egan, Timothy 256
Egypt — allocation of assistance 85
Ehrlich 19, 22, 29, 32, 59, 117, 118, 120,
  121, 123, 126, 128
Ehrlich & Ehrlich 19, 22, 23, 25, 27, 29,
  30, 36, 100, 117, 118, 206
EKD 223
El Salvador — allocation of assistance
  85
electricity supplies — Xishuangbanna,
  China 113
Elephant, Asiatic 112
Elephas maximus 112
emergent pressures stemming from
  population 17
Emlen 39
Emperor of Japan, The 209

employment opportunities 11
empowerment of women 236
–, men's & women's 2
energy 29
– consumption 237
– consumption necessary 239
–, efficient use of 238
– – – fertility 241–4
– from renewable resources and sources
  241
– levels, unsustainable 128
–, over- and mis-use 37
– – poverty 236
–, renewable 128
–, role of, in reducing growth-rate of
  human population 236
–, – –, in socio-economic development
  236
– use 128
'Engaged Buddhist' network 220
Engelman, Robert 47–77
Engels, Friedrich 95, 229, 230
England — demographic transition 285
Entwisle 244
environment, Earth's 5
–, perceptions of 134
–, physical 135
–, social 136
environmental accounting 124
– change 12
– constraints 101
– damage, commodities causing 123
– –, forms and costs to different countries
  of 262
– decline 25
– degradation 2, 19, 20, 121, 124
– – and poverty 125
– –, global 236
– –, population-associated 28
– – – poverty – high population 20
– –, strategies for reducing 36
– deterioration – poverty – population
  pressures 94
– devastation 53
– discontinuity 31
– disruptions 32
– equivalent of deficit financing 124
– ethics 208
– imbalances, increase of 55
– limits to population growth 59
– problems attributable to population
  growth 25–7
– – exacerbating population problems 27
– quality, declining 60
– refugees 32, 126, 176
– –, increase of 82
– resources, safeguard of 36–7
– sustainability 131

Printed in the United States
By Bookmasters